TRENDS IN GLOBAL RESEARCH IN WATER DESALINATION & DECONTAMINATION

By Mindy Levine

ISBN 978-1-57440-642-9

TABLE OF CONTENTS

Table of Contents

ABOUT THE AUTHOR

Associate Professor Mindy Levine joined the faculty at Ariel University (Israel) in the fall of 2019, where she is currently doing highly interdisciplinary organic chemistry research, teaching classes, and mentoring girls in science. Prior to her work at Ariel University, Mindy was a faculty member at the University of Rhode Island in the United States from 2010-2019. She received her B.A., M.A., and Ph.D. from Columbia University in New York City, and her doctoral research was focused on investigating the origin of homochirality. Following the completion of her Ph.D. in 2008, Mindy completed an NIH-funded postdoctoral fellowship at MIT in the laboratory of Professor Timothy Swager.

She is a strong advocate for the promotion of gender diversity in chemistry, through her work as a former chair of the Northeastern Section of the American Chemical Society; as the founder, coordinator, and leader of Chemistry Camp for Girls in Rhode Island; and as a mentor for middle school and high school girls. She has published more than 60 peer-reviewed publications and presented the results of her research in close to 200 presentations. In her spare time, Mindy enjoys spending time with her three children, eating ice cream, and conducting formal and informal science outreach.

WATER RESEARCH IN THE UNITED STATES

Introduction

Water research in the United States is funded by a variety of federal agencies, including the National Institutes of Health (NIH) through the National Institute for Environmental and Health Sciences (NIEHS); the National Science Foundation (NSF), predominantly through its CBET (Chemical, Bioengineering, Environmental, and Transport) Division; the Department of Energy (DOE), through individual investigator grants; and the Department of Interior (DOI), primarily through Title XVI ("The Water Reclamation and Reuse Act") and through the Water Desalination Act of 1996. A variety of private companies, either working in collaboration with a university or independently, also work in this area. Major players in this field in the United States are profiled below:

Brown University

A NIH-funded Superfund project at Brown University is funding researchers to study "Toxicant Exposures in Rhode Island: Past, Present, and Future," which includes a sub-focus on water purification.[1] Researchers at Brown University are partnering with state government officials to develop effective water testing techniques, with a particular focus on removing organic contaminants such as polyfluorinated and perfluorinated alkyl substances (PFAS).[2] This program has received over $10 million in funding through the NIEHS.[3]

Massachusetts Institute of Technology (MIT)

Significant effort at MIT is spent on water purification and desalination research. This effort includes the development of an electrochemically-driven method to remove micropollutants (via an individual research grant to T. Alan Hatton);[4] the development of xylem-based, low-cost water filters (via an individual research grant to Rohit Karnik);[5] and the development of photovoltaic-powered reversal desalination (through a Department of Interior grant to John H. Lienhard V).[6] MIT also has developed devices for water purification in developing countries, including in India, using technology developed by MIT professor Amos Winter and graduate student Natasha Wright.[7]

MIT also has an NSF-funded materials research center, which includes a focus on materials for water purification membranes, with more than $16 million in funding obtained to date.[8] There is also an Energy Initiative at MIT,[9] which includes research in the area of sustainability in water resources, including in purification and desalination of water. Finally, the Institute established the Abdul Latif Jameel Water & Food Systems Lab (J-WAFS), to coordinate water, energy, and food research across the institute.

Companies launched by MIT faculty and students also have a substantial focus on water purification and desalination. These include: (a) SiPure, a start-up founded by MIT students to purify textile wastewater;[10] (b) Infinite Cooling, a start-up that can provide a more effective way of cooling that limits water usage requirements;[11] and (c) Xconomy, a company that uses MIT-based technology to split water into hydrogen and oxygen and obtain energy from that process.[12]

National Aeronautics and Space Administration (NASA)

The National Aeronautics and Space Administration (NASA) is a big player in the water purification industry.

This work includes the following key components:

(a) Collaborations with local wastewater treatment companies to develop water purification technologies. For example, SageGuard Solutions (an Ohio-based company)[13] bought technology from the NASA Glenn Research Center to accomplish water purification via electrochemical means, at a cost of $50,000 up front plus 4% of gross sales.[14]

(b) Partnerships with non-profit organizations to bring water purification technologies to developing countries.[15] For example, a non-profit group, Concern for Kids,[16] worked with NASA researchers to bring water purification efforts to Iraq.[17]

(c) Grants to individual investigators to study the water cycle and water resources, which are given through the Science Mission Directorate (SMD) of NASA. Note that the NASA budget for FY2020 is $21 billion; of that, approximately 28% is budgeted for scientific research ($5.9 billion).[18]

New Jersey Institute of Technology (NJIT)

Research at NJIT in water research and purification includes an NSF-funded Industry-University Cooperative Research Center program for a center in "Membrane Science, Engineering and Technology (MAST), with Kamalesh Sirkar[19] and Boris Khusid[20] listed as co-PIs on the grant. $200,000 has been awarded to NJIT through this mechanism to date, which has resulted in numerous publications[21] and patents[22] based on the results of this research. An additional researcher at NJIT, Professor Linda Cummings,[23] has also received NSF funding for research in membrane filters for water purification, through the NSF GOALI (Grant Opportunities for Academic Liaison with Industry)[24] funding mechanism.

Northwestern University/ CycloPure

William Dichtel (Professor at Northwestern University)[25] founded CycloPure, a company that uses cyclodextrin-based materials for water purification.[26] CycloPure recently received $1 million in SBIR (Small Business Innovation Research) funding.[27] To date, CycloPure has one commercial product, a PFAS testing kit called "Water Test Kit Pro."[28] Dichtel himself has received substantial NSF funding to conduct research in the development of novel membranes for water purification research (through the CBET program) in the Engineering Division.[29]

University of California Berkeley

The University of California Berkeley (UC Berkeley) has a water research facility, termed the Berkeley Water Center (BWC),[30] which conducts substantial research in a variety of water-related areas including in purification of drinking water, responsible usage and disposal of wastewater, and management of natural water systems. The BWC is co-directed by Professors David Sedlak[31] and Isha Ray.[32] There are also substantial collaborations between the BWC and the Chinese Center for Disease Control and Prevention, as well as between the Berkeley Water Center (BWC) and the University of the Philippines (UPD) Marine Science Institute. Within UC Berkeley, the BWC is supported by the College of Natural Resources and the College of Engineering. The BWC receives substantial external funding from the National Institutes of Environmental and Health Sciences (NIEHS) as a Superfund Center (grant number: P42ES004705), with over $2 million in funding from the NIEHS received to date.

University of California Los Angeles

Professor Yoram Cohen at the University of California Los Angeles (UCLA)[33] has received funding from the Department of Defense for a grant entitled, "Novel Techniques for Energy-Optimal, Fault-Tolerant Desalination with Application to a Compact Modular RO System," which focuses on the development of novel desalination systems. Professor Cohen has published multiple papers on the topic of improved membrane filtration in desalination systems, as well as several patents. A selection of these papers[34] and patents[35] are included in the reference section of this report.

In addition, UCLA Professor Subramanian Iyer[36] has received $2.1 million in funding through the Department of Energy Small Business Innovation Research (SBIR) program for his company, named NRGTEK, INC.[37] The funding was to support company research in a solar-assisted seawater desalination program. Professor Iyer has filed several patents in the area of separations and osmosis processes, some of which are included in the reference section of this report.[38]

University of New Mexico

The University of New Mexico has received $5 million from the NSF for a research center entitled, "Phase II CREST Center for Water and the Environment."[39] The Center is run by Dr. Kerry J. Howe, a professor of civil engineering at the University of New Mexico,[40] and conducts research in the areas of watershed development, water treatment technologies, and research at the nexus of water and energy. There is a strong educational component to this research as well.

WATER RESEARCH IN AUSTRALIA

Introduction

The Australian Research Council has a significant interest in water purification and desalination due to limited availability of drinking water on the continent. The limited water availability worsened significantly during an extended drought period in Australia from 1997-2009,[41] which led the government to establish multiple desalination plants (more than 30 currently operating).[42] These plants are mostly government funded.[43] Research in water purification generally focuses on the removal of PFAS materials from the water supply.[44] Funding for this research has been provided to researchers at Adelaide University, Newcastle University, the University of Sydney, and the University of South Australia.[45] The research occurring at each institution is discussed in more detail below:

Adelaide University

The Water Research Centre at Adelaide University conducts research in a variety of water-related areas,[46] with a particular focus on developing methods to improve environmental water efficiency. The Centre is managed by Professor Justin Brookes,[47] in the Department of Ecology and Evolutionary Biology,[48] who has published research in a variety of high impact water-related research areas.[49]

Newcastle University

The Institute for Sustainability at Newcastle University conducts intensive research in a variety of water-related areas, including in the development of wastewater treatment technology and the development of improved methods for water harvesting technologies, including desalination efforts. Much of the water research is focused in the Department of Civil Engineering, with a particular focus the director of the Institute, Professor Philip McGowan.[50] Another institute at

Newcastle University, called BE:WISE, focuses on developing improvements in wastewater treatment technology.[51] This institute is led by principal investigator Dr Russell Davenport.[52] Both Professor McGowan[53] and Dr. Davenport[54] have published substantially in conservation-related research areas, with Dr. Davenport particularly focused on improvements in wastewater treatment technology.

University of South Australia

Water research at the University of South Australia is primarily in the "Natural and Built Environments Research Centre,"[55] which includes "Water and Natural Resources" as one of its four main research strands. The leader of this research strand is Associate Professor Md. Mizanur Rahman, in the Department of Geotechnical Engineering at the University of South Australia.[56] Publications from this strand include a substantial focus on wastewater treatment, management, and purification options.[57]

University of Sydney

Substantial research at the University of Sydney in the area of water, resources, and the environment occurs predominantly in the college of engineering at the university.[58] Research in the development of novel water treatment methods, including in wastewater treatment and desalination methods, is conducted by a broad range of faculty members at the university, including Professors John Kavanagh,[59] Yuan Chen,[60] Dianne Wiley,[61] and several other faculty members, all of whom have published extensively in this research area.

WATER RESEARCH IN CANADA

Introduction

Water research in Canada is motivated primarily by accelerating climate change, with researchers in this country acutely aware of how climate change is affecting ecosystems,[62] particularly in Northern Canada, and how the results of such climate change can have profound effects on the availability and distribution of water resources for the Canadian population.[63] Many such effects, including the prevalence of toxic algae in much of Canada's freshwater supply,[64] have already been realized. Efforts to develop methods for water purification and re-use, particularly in the area of wastewater treatment, have received substantial attention and funding from the Canadian government,[65] primarily in the form of research grants to universities but also through grants from private foundations, companies, and charities, including the Canada Water Resources Association.[66] Water research at many of the main universities is summarized below:

Carleton University

This university houses the Global Water Institute, which is home to more than 100 researchers studying challenges in water purification, conservation, and re-use.[67] The institute is led by Professor Banu Ormeci from the Department of Civil and Environmental Engineering,[68] who conducts research in wastewater treatment with implications in public health and environmental remediation. Professor Ormeci has multiple publications[69] and patents[70] related to this high-impact research. The research at the Institute is funded by a $4 million grant from the Jarislowsky Foundation,[71] as well as a variety of other funding sources.

McGill University

There are two main research centers for water at McGill University, the Water Innovation Lab[72] and the Integrated Water Resources Management Program.[73]

University of Saskatchewan

The Global Institute for Water Security is housed at the University of Saskatchewan,[74] with a particular focus on understanding and protecting water resources in the face of serious climate change. The Executive Director of the Institute, Professor Jay Famiglietti, is from the School of Environment and Sustainability, in the Department of Geography and Planning at the University of Saskatchewan.[75]

University of Toronto

The Institute for Water Innovation at the University of Toronto conducts research to address water challenges.[76] The director of the institute, Professor Vladimiros Papangelakis, has published on his substantial research efforts to improve water purification and desalination processes.[77]

University of Waterloo

The Water Institute, housed at the University of Waterloo, conducts high-impact research in the area of water purification and desalination.[78] More than 150 faculty members participate in this institute, which is run by water economist Professor Roy Brouwer.[79] Of particular interest is the research project in "Global Water Futures," which is part of the Water Institute[80] and is funded by a $77.8 million grant from the Canada First Research Excellence Fund.

WATER RESEARCH IN CHINA

Introduction

Desalination in China has been slower to develop than desalination in other countries, with a relatively recent push to develop desalination plants leading to 131 ocean desalination projects by the end of 2016. This new push has been driven by water shortages, which are particularly acute in certain geographic areas.[81] Nonetheless, significant technical challenges in achieving effective desalination and in using desalination to solve the issue of water security in China have been documented.[82] Water desalination, purification, and re-use efforts have occurred in private companies, government organizations, and research universities, many of which are reviewed in detail below.

College of Water Sciences, Beijing Normal University, Beijing

Beijing Key Lab of Urban Hydrological Cycle and Sponge City Technology[83] has conducted research in the area of urban flood mitigation, with research funded by the National Natural Science Foundation of China (grant number 51579007) focused on the feasibility of creating a "sponge city." This laboratory is part of Beijing Normal University College of Water Sciences (http://cws.bnu.edu.cn/). Research in the College of Water Sciences (CWS) is aimed primarily at enhancing water security for Chinese residents, specifically focused on urban water cycles, pollution mitigation, groundwater remediation, and seawater desalination. Funding for this research has come from the National Natural Science Foundation of China, and success in these research endeavors has been recognized through the awarding of prizes from the Ministry of Education, the Dayu Scientific and Technological Achievement Award, and others.

Fudan University, Shanghai, China

Fudan University is a public research university located in Shanghai.[84] Research in water purification and desalination at this university is conducted primarily in the Department of Environmental Science and Engineering,[85] with funding for this research coming from the National Natural Science Foundation, the National Nature Social Science Fund, and the Ministry of Science and Technology. Professor Xiang Liu works in this research area and has published 47 peer-reviewed publications related to water purification.[86]

Harbin Institute of Technology, Harbin, China

The School of the Environment at Harbin Institute of Technology (HIT) (http://sc.hit.edu.cn/10414/list.htm) has four featured research clusters listed on its website, and three of those four relate to water security and water purification. Harbin Institute of Technology hosted the Urban Water Science Forum in 2016. Key researchers at HIT include Professor Jun Ma, who is also the Associate Editor of the journal *Water Research*[87] and won a Sustainable Water Award in 2016. His work has been funded by the National Key Research and Development Program and by the Special Fund for School of Environment and Ecological Innovation of HIT. He is joined by Dr. Nan-Qi Ren, who is the Director of the State Key Lab of Urban Water Resource and Environment at HIT, and whose work focuses on water purification and pollutant removal.[88]

Sun Yat-sen University, Guangzhou, China

Primary work in this area is done by Professor C. M. Kao,[89] who is a Professor of Environmental Engineering who studies wastewater management and purification processes. Professor Kao has published his work in 60 journal articles to date on the topic of water purification. He is joined at this university by Professor Fangang Meng,[90] who works in the Research Center for Eco-Environmental Sciences at Sun Yat-sen University.

Tsinghua University, Beijing, China

Tsinghua University is one of the premier research universities in China.[91] Within this university, research in the Department of Environmental Sciences has focused on water purification and wastewater treatment efforts. This includes work by Professor Jianlong Wang of the Department of Environmental Sciences,[92] who has published more than 200 papers related to water purification research. Several of the listed research areas on the Department of Environment's website relate to water research, water purification, desalination, and wastewater treatment processes.[93]

University of Science and Technology of China, Hefei, China

Much of the research in water purification at this university has been spearheaded by Professor Han-Qing Yu,[94] who works in the Department of Chemistry and Materials Science,[95] and particularly in the CAS Key Laboratory of Urban Pollution Conversion.[96] Overall, solving the problem of water security in China has an extremely strong focus on treating urban areas and providing water to urban populations, due to the markedly higher percentage of the population that lives in cities in China compared to in other countries. Research in the Key Laboratory of Urban Pollution Conversion has been funded by a number of sources, including the National Science Foundation of China.

Wuhan University, Wuhan, China

Relevant research at this university[97] is conducted mostly in the Department of Environmental Science and Engineering, under the work of Professor Hui Zhang.[98] His work includes a focus on wastewater treatment and on pollutant removal, with a particular focus on using ultraviolet light to achieve targeted pollutant decomposition. Joining him in his research efforts is Professor Feng Wu[99] who works in the School of Resource and Environmental Sciences at Wuhan University.[100]

WATER RESEARCH IN FRANCE

French research in the area of water desalination has been significant, both in the industrial sector as well as in university-based research laboratories.

INDUSTRIAL SECTOR

Suez

In the industrial sector, Suez,[101] a utility company based in France, has done significant work in developing water desalination processes. Much of their technology is based in reverse osmosis for desalination processes. This reverse osmosis technology is used not only in France, but also in Barcelona to improve drinking capacity during a water shortage in 2008 and was able to dramatically increase the drinking water supply for the residents of the greater Catalonia area.[102]

Veolia

Veolia is a large, multinational company that has branches throughout the world. One branch of Veolia, called Sidem, is focused on desalination technology and the implementation of thermal-based desalination processes.[103] Sidem took part in the design, construction, and implementation of a hybrid desalination plant in the United Arab Emirates, called Fujairah 2, that was listed the sixth largest desalination plant in the world.[104] This plant uses both thermal energy and reverse osmosis technology to effectively provide 591,000 m^3 of pure drinking water per day.

ACADEMIC SECTOR

PSL Research University Paris (Paris Sciences et Lettres)

PSL has an environmental research center of excellence, termed the "PSL Environment."[105] This center is run by Dr. Serge Planes, and brings together researchers in a broad range of areas of environmental studies, including water desalination and water purification research. While the main research of Dr. Planes is in the area of marine animals and regulating the environment that surrounds such animals,[106] other researchers at the institute focus more heavily on water desalination and purification, including a particular focus on the development of artificial membranes for water purification purposes pioneered by Dr. Mihail Barboiu.[107,108]

Sorbonne University

Sorbonne University is really a collection of 11 different institutions,[109] with the UPMC (Pierre et Marie Curie University) particularly focused on science and medicine and UTC (Compiegne University of Technology) particularly focused on engineering. Research at the UPMC in the area of water purification and desalination includes a focus on membrane-based desalination processes, pioneered by Dr. Carmem-Lara de Oliveira Manes.[110] Water-related research at UTC is conducted by a variety of faculty members and their associated research groups, including Professor Andre Pauss,[111] whose work focuses primarily on wastewater treatment and associated water purification processes.[112]

Ecole Polytechnique

This relatively small university is still a major research site for science and technology, including in the area of water purification and reuse. In particular, the water-related research at this institution is conducted in the "LADHYX – Hydrodynamics" laboratory, which focuses on water harvesting and methods for water recycling. Whiel other research in sustainability is conducted in the molecular chemistry laboratories, for example, the key focus of the university in this area remains on identifying new water harvesting options and improving the performance of the options that are currently utilized.[113]

Université Paris-Sud

This university is one of the top universities in France (tied for 6th in a recent World University Ranking survey), with a strong focus on science and engineering. Universite Paris-Sud houses an interdisciplinary laboratory focused on environmental research, called "Centre international de recherche sur l'environnement et le développement" ("the international center for research on the environment and development").[114] One researcher associated with the center, Dr. David Makowski, focuses on the pollution of drinking water by pesticides and other organic pollutants,[115,116] and other researchers focus on the removal of such pollutants and on mechanisms to avoid the use of such pollutants from the outset.

WATER RESEARCH IN GERMANY

Introduction

Germany has a few desalination plants, which are primarily used to provide drinking water on the island of Helgoland, a small island found north of mainland Germany. This relatively small operation, however, is expected to increase in size dramatically as the population of Germany grows further and the demand for fresh drinking water outpaces the supply of water currently available.[117] Work in this area, like the work in France, takes place both in the industrial and in the academic sectors. Industrial sectors are marketing water treatment devices to combat notoriously 'hard' water in Germany, whereas academic research institutes are involved in a broad range of research areas related to water purification and desalination. Each of these sectors is discussed in turn.

INDUSTRIAL SECTOR

BestWater Germany[118]

This company sells water filtration devices directly to consumers, for use in their drinking water and water filtration systems, with technology that relies on reverse osmosis filters. This company was recently ranked #1 in the water purifiers market in Germany, with sales of home water purifiers expected to increase dramatically, surpassing $1 billion by 2023.[119] Much of this increase in direct to consumer sales is driven by the fact that German water is known to be "hard," (contain high mineral contents), which makes it less than ideal for some consumers to drink.[120] The increase in selling to big businesses has been driven by an increase in such business in Germany in recent years, as well as a desire to avoid the detrimental effects of hard water on expensive machinery (calcification, mineral deposition, clogging) and/or equipment parts.

Seccua[121]

Seccua is another German-based company that provides water filtration devices, both directly to consumers as well as to big businesses. This technology relies on the use of ultrafiltration membranes to achieve effective filtration. Compared to the reverse osmosis membranes used by BestWater, the ultrafiltration membranes have the potential to remove even more contamination from the water supply. Such improved performance, however, comes at an increased cost, and at this point, the BestWater filtration systems continue to outsell the filtration systems produced by Seccua. Both Seccua and BestWater Germany have strong international presences as well, which provides additional exposure and markedly increased additional potential markets for water purification sales.

ACADEMIC SECTOR

Karlsruhe University of Applied Sciences[122]

With the goal of providing an integrated center of excellence in water purification and water research, the Karlsruhe University of Applied Sciences signed an agreement in 2012 to partner with the following institutions: the Centre of Environmental and Health Engineering (CEHE) of the University of Surrey, the Water Technology Center of the German DVGW (German Association for Gas and Water Technology), The Heinrich Sontheimer Laboratory (HSL) and the Karlsruhe Municipal Services (SWK).[123] The water treatment research at the university is conducted primarily in the Institute of Refrigeration, Air-Conditioning, and Environmental Engineering.[124] The director of the water research center, Dr. Josef Klinger, has conducted substantial research in measuring pollutants in groundwater and drinking water using a variety of analytical techniques.[125,126]

Technische Universität Berlin (TU Berlin)

TU Berlin has an extremely strong Department of Water Engineering, located at the Campus El Gouna of TU Berlin.[127] The stated goal of the department is to focus on an "integrated and interdisciplinary" approach to addressing water shortages and enhancing water security for a global population. Water-related research in this department has been conducted by a number of faculty members, including Dr. Nikolaus Tzoupanos, whose research focuses on wastewater purification processes[128] and on mitigation of current levels of environmental pollution.[129] Projects undertaken through this center include "Decentralized treatment wetlands for sustainable water management in rural and remote areas of semi-arid regions," and "Biofilm-Hybrid-Process for Wastewater Treatment in North African Climate: Conventional Carrier vs Recycling Material for Capacity Increase."

Karlsruhe Institute of Technology

This institute has a Chair of Water Chemistry and Water Technology,[130] and research in this area focuses on analysis of water quality as well as the development of new methods for water treatment. The institute is currently led by Professor Harald Horn,[131] whose research focuses primarily on the development of new methods for aqueous pollutant removal,[132] and is co-chaired by Dr. Gudrun Abbt-Braun, whose research focuses on the development of water treatment membranes and reverse osmosis processes.[133] Moreover, the institute also has a strong focus on academic training, with students in the institute learning about aquatic systems and wastewater treatment processes.

WATER RESEARCH IN INDIA

Introduction

The large and rapidly increasing population in India means that there is a continued need to provide safe drinking water for the population. Increasingly, government officials in India are turning towards desalination and wastewater treatment to provide improved drinking water. The largest desalination plant in India, called the Minjur Desalination Plant,[134] relies on reverse osmosis to provide drinking water to the city of Chennai (Madras), the 6th most populated city in India. Other desalination plants in India, including one in Tamil Nadu, use solar power to enable effective desalination,[135] although concerns about the large-scale feasibility of solar power remain.[136] Significant research in this area has been conducted in the government, industrial, academic, and international sectors.[137] Each of these areas will be discussed in turn.

GOVERNMENT SECTOR

Of the two major desalination plants in India, one is run by private industry and one is run by a local government entity. The Nemmeli Desalination Plant located near Nemmeli, Chennai, is owned by the Chennai Metrowater Supply and Sewage Board, a local government agency,[138] and is used to provide drinking water to residents of the city of Chennai. The mechanism of desalination employed by the plant is ultrafiltration, which provides a mechanism to remove all impurities from the drinking water, at an estimated cost of 36 Indian rupees (approximately $0.48) per kiloliter of potable water produced.

INDUSTRIAL SECTOR

Private industry runs the largest desalination plant in India, the Minjur desalination plant (mentioned above). In addition to the desalination efforts, substantial industrial work has led to the development, manufacture, and direct to consumer sales of water filtration systems. Companies such as Xiaomi,[139] Livpure,[140] and Kent,[141] have developed water filters and have published articles explaining why the use of such filters is critical in India, due to increased water pollution and lack of regulation around water quality in the country. Recent reports estimate that approximately 70% of India's water is contaminated with pollutants,[142] and substantial water-borne illness has resulted.[143]

ACADEMIC SECTOR

Indian Institute of Science (IISc), Bangalore, India[144]

The IISc houses an Interdisciplinary Centre for Water Research, which was founded in 2015 to address challenges in water purification, urban water systems, hydrology, and desalination.[145] Ten faculty members are associated with the research center, which is led by Professor P. P. Mujumdar from the Department of Civil Engineering.[146] Professor Mujumdar has published extensively on the effects of climate change on water management and water control, highlighting the increasing risks of contamination introduced via accelerated climate change.[147][148][149] Significant media coverage of the work done at this center includes a spotlight on how research is being used to predict rain and tackle floods,[150] as well as a general article focused on India's water problem.[151]

Vishwakarma University, Pune, India[152]

This university provides a course in water and wastewater treatment, which can be taken either full time (for 1.5 months) or part-time (for 3 months).[153] This course is an outgrowth of the University's Water Quality Centre of Excellence,[154] run by center director Dr. Dhananjay Bhatkhande and center deputy director Dr. Shraddha Khamparia. Both the center director and deputy director have published in areas related to water purification and water resource allocation, with Dr. Bhatkhande focusing on wastewater treatment[155] and Dr. Khamparia focusing on the removal of pollutants in treated water.[156] The center has a strong research and laboratory component, and also runs community services through testing water samples sent by members of the community. Initial and primary funding for the center comes from the Wilo Foundation, a private, German-based foundation.[157]

INTERNATIONAL SECTOR

TU Delft

Researchers at TU Delft (in the Netherlands) have initiated a research program to improve water availability and the safety of the water supply in India.[158] These efforts have been led by TU Delft professor Merle de Kreuk, whose research focuses on understanding microbial contamination in water[159] and on wastewater management technologies,[160] and Ralph Lindeboom, whose research focuses on resource management and allocation for long-term Space missions.[161] Of note, the TU Delft water research program was recently ranked one of the top water research centers world-wide.[162]

Hamburg University of Technology

Researchers at the Hamburg University of Technology (in Germany) have established an Institute of Wastewater Management and Water Protection.[163] Researchers at this institute work in a number of geographic areas, including in India, and focus on three main research thrusts: (a) rural revival and restoration engineering; (b) resources oriented sanitation systems; and (c) bioresource management.[164] Leaders of all three research thrusts have extensive experience publishing in the area of water purification,[165] and salt water desalination technology.[166]

WATER RESEARCH IN ISRAEL

Introduction

Israel as a country has a severe shortage of natural water resources.[167] Starting in 2000, large-scale government sponsored desalination efforts began, with the current guidelines for desalination set forth in the government master plan known as "Tama 34b2/2." The leader of desalination efforts in Israel is the company IDE Technologies,[168] which operates the Sorek desalination plant, the largest seawater desalination plant in the world.[169] Most of the large-scale desalination plants in Israel are privately funded, although bond issues, government grants, and soft loans are also used.

Research in this area is funded by the Israel Innovation Authority,[170] which funds both individual researcher grants as well as collaborative research with foreign institutions; the Israel Science Foundation (ISF), which funds fundamental research that is often relevant to water purification and desalination efforts;[171] and the Ministry of Science and Technology.[172] Moreover, within each institution of higher education, innovation offices provide funding for research that has significant commercial potential, which often includes research in the area of water desalination and/or water purification.

University research in this area is being conducted at the Technion Institute, at Ben Gurion University, and at Hebrew University of Jerusalem. The research occurring at each institution is discussed in more detail below:

Ben Gurion University

Water research at Ben Gurion University (BGU) is done primarily at the Zuckerberg Institute for Water Research,[173] which is directed by Professor Amit Gross.[174] Research at the Zuckerberg Institute focuses on environmental hydrology and microbiology, as well as the areas of desalination and water treatment research. The existence of a pilot plant for desalination at one of BGU's campuses, in Sde Boker, provides an opportunity for practical research and development of novel desalination methods. Research at this institute is funded by the Israeli government as well as by private funding.

Hebrew University of Jerusalem

Research at the Hebrew University of Jerusalem in the area of water purification and desalination occurs primarily in the Department of Soil and Water Sciences.[175] The head of the department is Associate Professor Yael Mishael,[176] who conducts research in the area of clay-based filters for water purification systems.[177] Significant research efforts in the department also focus on methods for improved agricultural irrigation that minimize the use of water and allow for improved conservation of limited water resources.

Technion Institute of Technology

The Technion is home to the Grand Water Research Institute (GWRI),[178] which includes a substantial focus on water treatment and desalination research. This institute includes faculty members from multiple departments and colleges throughout the Institute, including from Engineering (chemistry, civil, environmental, and materials engineering departments), from the Exact Sciences (chemistry and biology), and many other departments. The director of the GWRI is Professor Ori Lahav from the Civil and Environmental Engineering Department at the Technion.[179]

WATER RESEARCH IN ITALY

Introduction

There is significant rainfall in Italy; however, that rainfall is insufficient to meet the drinking water needs of the entire Italian population. This is true in all parts of the country, but is a particularly acute problem in southern regions of Italy, where only 30% of the drinking water need is met through rainfall.[180] Of note, tap water in the country is generally considered safe to drink, although high levels of minerals (especially calcium) in the water provide a "hardness" that may affect consumers and/or tourists and lead to an enhanced market for water treatment options. As such, desalination efforts in Italy comprise a necessary and increasing industry, mostly focused on private industry and on university-led research efforts in the area of water desalination, water purification, and wastewater treatment. Each of these sectors are discussed herein.

INDUSTRIAL SECTOR

Sarlux IGCC Power Plant[181]

This power plant, located in Sardinia, Italy, is one of the largest such plants in Europe. It combines gasification of residual oils (i.e. converting waste oil into useful gasoline) with a variety of other efforts, including wastewater pre-treatment and desalination. The plant was commissioned in 1999, and its construction was financed by a $1.2 billion loan.[182] At this time it is being operated by IDE Technologies, an Israel-based company,[183] and is capable of providing an estimated 17,280 m^3/day of clean water.

Acconia Desalination[184]

The largest desalination plant in the Mediterranean geographic area was completed in 2018, and is located in Sardinia, Italy. By contract, it is scheduled to be operated for six years by ACCIONA

Agua, and to produce 12,000 m^3/day of high quality, demineralized water. The plant uses both ultrafiltration and reverse osmosis technology (as well as other processes) in order to enable effective desalination to occur. The operating company, ACCIONA Agua, has a large number of other water treatment processes scattered throughout Italy.

European Desalination Society (EDS)[185]

This society includes members in academia, industry, and government. It is headquartered in Rome, Italy, and has members throughout the European continent. Activities of the society include desalination, as well as a broad spectrum of research and applied activities in the area of water purification and water treatment.

Other coupled desalination plants

In addition to the Sarluxx power plant listed above, there are a number of other coupled plants in Italy, defined as plants that combine desalination with other industrial processes. Such coupling include the coupling of reverse osmosis-powered desalination with concentrating solar power;[186] coupling solid waste treatment with a multi-stage flash desalination plant;[187] and the coupling of multi-flash desalination with power cogeneration at a third site in Italy.[188]'

MEMBRANE

In addition to the desalination work done by a variety of private industrial companies, there are a large number of industrial companies in Italy working in the general area of water purification. One example is MEMBRANE, a company that creates filters and other devices to facilitate water purification.[189] The company produces a broad variety of membranes, including those used for reverse osmosis, those used for water desalination, and those that rely on ultrafiltration. The company is based in Milan, Italy, but has offices in the United Arab Emirates, in the United States, and in Malta.

ACADEMIC SECTOR

Universita Di Bologna

Work at the University of Bologna in water research is concentrated in the department of civil, chemical, environmental, and materials engineering. In particular, research by Associate Professor Valentina Ciriello has focused on the study of natural water systems[190] and on the management of the water supply in geographic regions throughout Italy.[191,192]

Politecnico Di Milano

The Department of Civil and Environmental Engineering (DICA) at the Politecnico Milano university has a research focus on water science and engineering.[193] Eleven faculty members at the university conduct research in this area, with a particular focus on understanding water resources management under accelerated climate change[194] and on managing water resources in urban environments.[195]

WATER RESEARCH IN JAPAN

Introduction

Although there is not a noticeable shortage of drinking water in Japan, with most of the water needs of the population met through rainfall or other natural resources, Japan has nonetheless developed pioneering technology in the area of desalination. The substantial access of Japanese people to seawater (as Japan is made up of a series of islands) means that the ability to use seawater for practical purposes can have substantial economic and other advantages. Methods to achieve effective desalination have been implemented by privately owned companies, and research into the development of novel water treatment technologies is ongoing at many of the top national universities. Each of these sectors (industrial and academic) is discussed in turn.

INDUSTRIAL SECTOR

Kyowakiden Industry Co., Ltd.[196]

This company was established in the immediate aftermath of World War II, in Nagasaki, and has as its stated goal the ability to provide resources to cities that are recovering from war. This goal includes facilitating water purification and the delivery of clean water to local residents, and for this company, that goal includes using reverse osmosis technology to accomplish effective desalination. In addition to its work in Japan, where the company is headquartered, Kyowakiden also works in a number of developing countries in Asia, including in developing villages in Indonesia where it is able to provide safe drinking water to residents.

Fukuoka District Water Works Agency[197]

The desalination plant run by this agency is the largest seawater desalination plant in Japan, and it provides drinking water to residents of Fukuoka City, a city that is often plagued by droughts and other water-related shortages. The agency that runs the desalination plant is part of the local municipal government and relies primarily on reverse osmosis and on ultrafiltration membranes to accomplish efficient desalination. Membranes for the desalination process are provided by an outside company, Toyobo Co., Ltd.[198]

Nitto Denko

This company is not responsible for desalination, per se, but is responsible for the manufacture of reverse osmosis membranes that are often used in desalination facilities. This publicly traded company is based in Japan and was founded in 1918. It has a broad product portfolio, of which reverse osmosis membranes represent a significant fraction.[199]

ACADEMIC SECTOR

Japan Water Research Center[200]

This water research center is a nonprofit organization, rather than a truly 'academic' one, whose stated goal is to promote healthy and safe drinking water to Japanese residents. To that end, the organization has been extremely successful in advocating for and implementing filtration technology to improve the quality of drinking water, as well as in hosting annual conferences that bring together leading researchers in the water purification and water treatment research areas.

Kyoto University[201]

Kyoto University, within its Department of Civil and Earth Resources Engineering, has significant research focused on hydrology, water purification, and water resources management. The work in this area is being executed by a large number of faculty members, including Professor Yasuto Tachikawa,[202] Professor Yutaka Ichikawa,[203] and Professor Kazuaki Yorozu.[204] Particular areas of research interest include analysis and numerical modeling of hydrologic processes, the impact of climate change, and the ability to control, manage, and conserve water resources.

Tokyo Metropolitan University[205]

Within Tokyo Metropolitan University, work in the department of civil and environmental engineering includes a strong focus on hydrology and water resources. Such research is being led by two faculty members, Professor Akira Kawamura and Assistant Professor Hideo Amaguchi. Professor Kawamura's research focuses on analyzing hydrogeochemical data of groundwater,[206] and Assistant Professor Amaguchi's work has a similar focus on groundwater monitoring and spatiotemporal classification.[207]

Kobe University[208]

This university houses a Lab of Water Environmental Research,[209] which conducts research focused on rainfall-runoff models, the efficient use of limited water resources in a broad geographic area, and on the development of ways to improve and effectively monitor water quality. Primary staff members associated with this research endeavor include Professor Haruya Tanakamaru and Associate Professor Akio Tada. Professor Tanakamaru's research focus is in the area of water management, with more than 200 publications in this research area,[210] and Associate Professor Tada's research focus is similarly focused on managing limited water resources.[211]

WATER RESEARCH IN SAUDI ARABIA

Introduction

Much like Australia and Israel, Saudi Arabia is a country that is lacking in natural water resources, and it relies heavily on desalination to obtain sufficient and safe drinking water for its population. Of note, recent reports indicate that Saudi Arabia leads the world in both the consumption and the production of desalinated water.[212] Desalination in the country occurs at 27 desalination plants, which together produce 3.3 million m^3/day of safe drinking water. Research into water desalination and water purification occurs at a variety of institutions throughout the country, including those that are profiled below:

Academic institutions

King Fahd University of Petroleum and Minerals

(KFUPM)[213] – A variety of departments at KFUPM are actively conducting research in the area of water desalination and water purification procedures. In the Department of Civil and Environmental Engineering, there is a Water Resources and Environmental Engineering Lab.[214] Through the resources present in this laboratory, researchers are conducting funded research in the area of wastewater purification and treatment using photocatalytic and advanced oxidative methods (led by Professors Muhammad S. Vohra and Mohammed S. Al-Suwaiyan); research on the effects of climate change on drinking water and how decision making regarding drinking water depends on a variety of complex and co-existing criteria (led by Professor Muhammad H. Essa); and research on the management of sustainable water processes (led by Professor Shakhawat Chowdhury). Each of these faculty members have published extensively on this topic, and representative publications from the aforementioned faculty members are included below:

Representative publications from Muhammad S. Vohra:

1. Kedir, A. J.; Vohra, M. S. Recent Developments in Aqueous Phase Bromate (BrO3-) Species Treatment Using Adsorption, Reduction, and Ion-Exchange Processes. *Desalination Water Treatment* **2017**, *80*, 255-267.
2. Kazeem, T. S.; Labaran, B. A.; Ahmed, H.-u.-R.; Mohammed, T.; Essa, M. H.; Al-Suwaiyan, M. S.; Vohra, M. S. Treatment of Aqueous Selenocyanate Anions Using Electrocoagulation. *Int. J. Electrochem. Sci.* **2019**, *14*, 10538-10564.

Representative publications from Mohammed S. Al-Suwaiyan:

1. Ishaq, A. M.; Al-Suwaiyan, M. S.; Khan, A. A. Quality Changes During Artificial Recharge of Wastewater Effluents. *J. Environ. Sci. Engineering* **2010**, *4*, 14-21.
2. Al-Suwaiyan, M. S. Controlling Groundwater Pollution from Petroleum Products Leaks. *WIT Trans. Ecol. Environ.* **2010**, *132*, 91-99.

Representative publications from Muhammad H. Essa:

1. Farooq, S.; Nakhla, G. F.; Essa, M. H. Simulation of In-Situ Bioremediation of Phenol Contaminated Sandy Aquifers. 2. Effect of Phenol Concentration. *Water Air Soil Pollution* **1996**, *87*, 283-295.
2. Saleem, M.; Essa, M. H.; Chakrabarti, M. H.; Low, C. T. J.; Hajimolana, S. A.; Hussain, M. A.; Hashim, M. A.; Islam, M. S. An Economical Means for Remediating a Multiple-Metal Contaminated Soil Using Electro-Kinetic Technology Under Natural Conditions. *Int. J. Electrochem. Sci.* **2017**, *7*, 7809-7822.

Representative publications from Shakhawat Chowdhury:

1. Chowdhury, I. R.; Chowdhury, S.; Al-Suwaiyan, M. S. Human Exposure and Risk of Trihalomethanes During Continuous Showering Events. *Sci. Total Environ.* **2020**, *701*, 134521.

2. Chowdhury, S.; Mazumder, M. A. J.; Alhooshani, K.; Al-Suwaiyan, M. S. Reduction of DBPs in Synthetic Water by Indoor Techniques and its Implications on Exposure and Health Risk. *Sci. Total Environ.* **2019**, *691*, 621-630.

In the Department of Chemistry, research in the area of polymers, materials, and nanoscience includes significant research on water treatment processes.[215] In particular, research in the department focuses on the development of novel polymers as resins for water treatment (led by Professor Shaikh Asrof Ali);[216] and the development of novel materials for effective wastewater treatment (led by Professor Othman Charles Al-Hamouz).[217] Each of these faculty members have published extensively in this research area, and representative examples of their publications are shown below:

Representative publications from Shaikh Asrof Ali:

1. Ali, S. A.; Rachman, I. B.; Saleh, T. A. Simultaneous Trapping of Cr(III) and Organic Dyes by a pH-Responsive Resin Containing Zwitterionic Aminomethylphosphonate Ligands and Hydrophobic Pendants. *Chem. Engineering J.* **2017**, *330*, 663-674.
2. Saleh, T. A.; Muhammad, A. M.; Tawabini, B.; Ali, S. A. Aminomethylphosphonate Chelating Ligand and Octadecyl Alkyl Chain in a Resin for Simultaneous Removal of Co(II) Ions and Organic Contaminants. *J. Chem. Engineering Data* **2016**, *61*, 3377-3385.

Representative publications from Othman Charles Al-Hamouz:

1. Albakri, M. A.; Abdelnaby, M. M.; Saleh, T. A.; Al Hamouz, O. C. S. New Series of Benzene-1,3,5-Triamine Based Cross-Linked Polyamines and Polyamine/CNT Composites for Lead Ion Removal from Aqueous Solutions. *Chem. Engineering J.* **2018**, *333*, 76-84.

2. Khalid, A.; Al-Juhani, A. A.; Al-Hamouz, O. C.; Laoui, T.; Khan, Z.; Atieh, M. A. Preparation and Properties of Nanocomposite Polysulfone/Multi-Walled Carbon Nanotubes Membranes for Desalination. *Desalination* **2015**, *367*, 134-144.

King Abdullah Institute of Science and Technology in Saudi Arabia (KAUST)[218]

This institute, located in Thuwai, Saudi Arabia, was established in 2009 and is a major research university today. Research at KAUST in the area of water desalination and water treatment takes place largely at the KAUST Water Desalination and Reuse Center.[219] Research at the center focuses on the development of innovations in desalination treatment, in wastewater treatment and reuse, in the development of novel membranes for effective water management, and in the use of sustainable water technologies for industry and agriculture. The center is led by Professor Johannes Vrouwenvelder (Director)[220] and by Professor Matthew McCabe (Associate Director).[221] Both of these faculty members have published extensively on results of their water treatment research, and representative publications are included below:

Representative publications from Johannes Vrouwenvelder:

1. Kerdi, S.; Qamar, A.; Alpatova, A.; Vrouwenvelder, J. S.; Ghaffour, N. Membrane Filtration Performance Enhancement and Biofouling Mitigation Using Symmetric Spacers with Helical Filaments. *Desalination* **2020**, *484*, 114454.
2. Farhat, N.; Kim, L. H.; Vrouwenvelder, J. S. Online Characterization of Bacterial Processes in Drinking Water Systems. *Clean Water* **2020**, *3*, 16.

Representative publications from Matthew McCabe:

1. Parkes, S. D.; McCabe, M. F.; Griffiths, A. D.; Wang, L.; Chambers, S.; Ershadi, A.; Williams, A. G.; Strauss, J.; Element, A. Response of Water Vapour D-Excess to Land-Atmosphere Interactions in a Semi-Arid Environment. *Hydrology Earth System Sci.* **2017**, *21*, 533-548.

2. Zhao, L.; Xiao, H.; Zhou, J.; Wang, L.; Cheng, G.; Zhou, M.; Yin, L.; McCabe, M. F. Detailed Assessment of Isotope Ratio Infrared Spectroscopy and Isotope Ratio Mass Spectrometry for the Stable Isotope Analysis of Plant and Soil Waters. *Rapid Commun. Mass Spectrometry* **2011**, *25*, 3071-3082.

Private companies

ACCIONA[222]

– This is a Spanish conglomerate that operates in a large number of countries, including in Saudi Arabia. It was the recent recipient of funding from the SWCC (vide infra) to develop the Al Khobar 2 desalination plant on the east cost of Saudi Arabia, through an estimated $500 million contract.[223] This facility will accomplish effective desalination primarily through the use of reverse osmosis technology. Of note, ACCIONA has also received significant funding from the SWCC to develop other desalination plants in Saudi Arabia, including the Shuqaiq3 desalination plant, the Al Khobar 1 desalination plant, and the Al Jubail RO4 desalination plant.

Abengoa[224] -

Another Spanish-speaking company, Abengoa, has received significant funding to develop desalination plants in Saudi Arabia.[225] Of note, this funding also derives from the SWCC, in its capacity to provide contracts for desalination plant development.

Government agencies

Saline Water Conversion Corporation (SWCC)[226] –

This is an extremely large publicly-owned company that operates both desalination plants and power stations throughout Saudi Arabia. Overall, the SWCC is responsible for 20% of desalination production world-wide. It operates in 30 plants throughout Saudi Arabia, and grants contracts and other funding to a variety of private agencies and institutions of higher education to continue to develop highly effective water desalination treatment procedures.

WATER RESEARCH IN SOUTH KOREA

Introduction

South Korea has a number of challenges that it must overcome in order to provide safe drinking water to its more than 51 million residents.[227] These include a limited supply of freshwater from naturally occurring sources,[228] the fact that although rainfall is abundant it is extremely unevenly distributed,[229] and the fact that there is significant environmental pollution in general, and water pollution in particular.[230] Addressing these water challenges has required research and development in the area of water purification, wastewater treatment and reuse, and seawater desalination. This research is occurring in academic institutions, private companies, and government agencies. Each of these areas are discussed in turn.

Research at academic institutions

Seoul National University

(located in Seoul, South Korea)[231] – In the Department of Civil, Architectural, and Environmental Engineering, Professor Mooyoung Han[232] is conducting research in rainwater harvesting methods. This work is being done primarily in the Rainwater Research Center (part of Seoul National University). The rainwater harvesting methods developed by the center have been highly successful both in South Korea and in nearby countries, in which researchers have used the rainwater harvesting methods to provide clean drinking water to Vanuatu, a Pacific island country.

Two sample publications by Professor Han on the topic of rainwater harvesting are listed below:

1. Tran, S. H.; Dang, Huyen T. T.; Dao, D. A.; Nguyen, V.-A.; Nguyen, L. T.; Nguyen, V.-A.; Han, M. On-Site Rainwater Harvesting and Treatment for Drinking Water Supply: Assessment of Cost and Technical Issues. *Environ. Sci. Pollution Res.* **2020**, *Ahead of Print*; DOI: 10.1007/s11356-020-07977-0.

2. Lee, M.; Kim, M.; Kim, Y.; Han, M. Consideration of Rainwater Quality Parameters for Drinking Purposes: A Case Study in Rural Vietnam. *J. Environ. Management* **2017**, *200*, 400-406.

Korea University

The School of Civil, Environmental, and Architectural Engineering,[233] located in Korea University (a private university in Seoul, South Korea),[234] is focused on a variety of research areas including those related to water treatment and water supply management. The faculty members most focused on water treatment methods include Professor Joong Hon Kim[235] and Professor Chulsang Yoo.[236] Representative publications are included below.

Sample publications from Professor Joong Hon Kim:

1. Seo, Y.; Kim, J.; Seo, U. Eco-Friendly Snow Melting Systems Developed for Modern Expressways. *J. Testing Evaluation* **2019**, *47*, JTE20180161.
2. Kim, H. K.; Kim, J. H.; Lee, B. C.; Yu, S. J.; Kim, H. J. Occurrence of Perchlorate in Drinking Water Sources in Korea. *Water Sci. Technol: Water Supply* **2009**, *9*, 133-139.

Sample publications from Professor Chulsang Yoo:

1. Yoo, C.; Ku, J. M.; Jun, C.; Zhu, J. H. Simulation of Infiltration Facilities Using the SEEP /W Model and Quantification of Flood Runoff Reduction Effect by the Decrease in CN. *Water Sci. Technol.* **2016**, *74*, 118-129.
2. Lee, J.; Pak, G.; Yoo, C.; Kim, S.; Yoon, J. Effects of Land Use Change and Water Reuse Options on Urban Water Cycle. *J. Environ. Sci.* **2010**, *22*, 923-928.

Research at private companies

Posco E&C[237] -

This is a global construction company, based in South Korea, but with offices and operations throughout East Asia, Saudi Arabia, and South and Central America. A significant portion of their operations in South Korea focus on wastewater treatment and management,[238] with an eye towards the development of sustainable and cost-effective processes. Recent work includes the development of a decentralized water supply network (with greater local control)[239] as well as the development of filtration technology for effective water purification.

2. Large scale desalination plants in South Korea are almost always operated by private businesses and companies. In one example, a large-scale reverse osmosis-based desalination plant was built near Busan, South Korea by Doosan Heavy Industries and Construction.[240,241] Although that plant was temporarily shut down due to concerns about the proximity of the plant to a nuclear power complex,[242] it was re-opened in 2019.[243]

Research at government agencies

K-Water[244]

This is a state-run organization, controlled by CEO Jae-Hyeon Park, which is based in South Korea. The company is working to develop desalination plants to provide effective drinking water. Although in many countries, large scale desalination plants have been developed, the desalination plants in South Korea, including those developed and implemented by K-Water, tend to operate on a much smaller scale. Small-scale desalination plants in general can provide good flexibility in terms of identifying locations for the plant construction, but suffer from significant challenges related to the inability to capitalize on economies of scale and to develop truly cost-effective desalination methods.[245]

WATER RESEARCH IN SWITZERLAND

Introduction

Switzerland is a country with significant water resources. These resources include significant rainwater (an average of 60 billion m^3 per year); snow pack and the associated melting of the snow (approximately 5 billion m^3 per year, all in the spring); ice caps; aquifers; and natural lakes.[246] Of note, the average available water per person per year is 5100 m^3, which is three times the global average of water availability. Nonetheless, vulnerabilities in the water supply remain, and generally focus on the unevenness of rainfall among the various geographic areas, as well as the fact that climate change is expected to significantly affect the availability of water from melting snow and ice caps, with an initial increase in water availability expected, followed by a significant decline. As such, the need for developing systems to analyze and purify drinking water, including through desalination and wastewater treatment, remains a high priority.

Research at academic institutions

Universite de Neuchatel[247]

This university is located in the French-speaking city of Neuchatel, in Switzerland. As part of the University, there is a highly productive laboratory, called the "Centre for Hydrogeology and Geothermics (CHYN),"[248] which conducts substantial water-related research. In particular, Professor Daniel Hunkeler[249] is studying the composition and purification of groundwater, with an eye towards using such water as part of a safe drinking water supply. He has published extensively on this topic, and some representative publications are shown below:

1. Hintze, S.; Glauser, G.; Hunkeler, D. Influence of Surface Water - Groundwater Interactions on the Spatial Distribution of Pesticide Metabolites in Groundwater. *Sci. Total Environ.* **2020**, *733*, 139109.

2. Melsbach, A.; Torrento, C.; Ponsin, V.; Bolotin, J.; Lachat, L.; Prasuhn, V.; Hofstetter, T. B.; Hunkeler, D.; Elsner, M. Dual-Element Isotope Analysis of Desphenylchloridazon to Investigate Its Environmental Fate in a Systematic Field Study: A Long-Term Lysimeter Experiment. *Environ. Sci. Technol.* **2020**, *54*, 3929-3939.

Swiss Federal Institute of Aquatic Science and Technology (Eawag)[250]

This institute is located within the ETH domain,[251] and has been ranked as one of the leading water research institutes in the world.[252] Relevant research at this institute includes the research on drinking water (both characterization of contaminants and purification of the water from those contaminants);[253] wastewater (including effective treatment strategies to minimize pollution and maximize potential for reuse);[254] and pollutants (including both the characterization of pollutants as well as methods to remove those pollutants).[255] Drinking water research is led by Professor Michael Berg; urban water management research is led by Professor Max Maurer, surface water research is led by Professor Carsten Schubert, and wastewater and solid waste management is led by Professor Christoph Luthi. Each of these faculty members has published extensively about his research, and representative publications from each faculty member are listed below.

Representative publications from Professor Michael Berg:

1. Podgorski, J.; Berg, M. Global Threat of Arsenic in Groundwater. *Science* **2020**, *368*, 845-850.
2. Suess, E.; Berg, M.; Bouchet, S.; Cayo, L.; Hug, S. J.; Kaegi, R.; Voegelin, A.; Winkel, L. H. E.; Tessier, E.; Amouroux, D.; Buser, A. M. Mercury Loads and Fluxes from Wastewater: A Nationwide Survey in Switzerland. *Water Res.* **2020**, *175*, 115708.

Representative publications from Professor Max Maurer:

1. Hoffmann, S.; Feldmann, U.; Bach, P. M.; Binz, C.; Farrelly, M.; Frantzeskaki, N.; Hiessl, H.; Inauen, J.; Larsen, T. A.; Lienert, J.; Londong, J.; Luthi, C.; Maurer, M.; Mitchell, C.; Morgenroth, E.; Nelson, K. L.; Scholten, L.; Truffer, B.; Udert, K. M. A Research Agenda for the Future of Urban Water Management: Exploring the Potential of Nongrid, Small-Grid, and Hybrid Solutions. *Environ. Sci. Technol.* **2020**, *54*, 5312-5322.
2. Mutzner, L.; Vermeirssen, E. L. M.; Mangold, S.; Maurer, M.; Scheidegger, A.; Singer, H.; Booij, K.; Ort, C. Passive Samplers to Quantify Micropollutants in Sewer Overflows: Accumulation Behaviour and Field Validation for Short Pollution Events. *Water Res.* **2019**, *160*, 350-360.

Representative publications from Professor Carsten Schubert:

1. Tomonaga, Y.; Brennwald, M. S.; Livingstone, D. M.; Kwiecien, O.; Randlett, M.-E.; Stockhecke, M.; Unwin, K.; Anselmetti, F. S.; Beer, J.; Haug, G. H.; Schubert, C. J.; Sturm, M.; Kipfer, R. Porewater Salinity Reveals Past Lake-Level Changes in Lake Van, the Earth's Largest Soda Lake. *Sci. Reports* **2017**, *7*, 1-10.
2. Ladd, S. N.; Nelson, D. B.; Schubert, C. J.; Dubois, N. Lipid Compound Classes Display Diverging Hydrogen Isotope Responses in Lakes Along a Nutrient Gradient. *Geochim. Cosmochim. Acta* **2018**, *237*, 103-119.

Representative publications from Professor Christoph Luthi:

1. O'Keefe, M.; Messmer, U.; Luthi, C.; Tobias, R. Slum Inhabitants' Perceptions and Decision-Making Processes Related to an Innovative Sanitation Service: Evaluating the Blue Diversion Toilet in Kampala (Uganda). *Int. J. Environ. Health Res.* **2015**, *25*, 670-684.
2. Tilley, E.; Strande, L.; Luthi, C.; Mosler, H.-J.; Udert, K. M.; Gebauer, H.; Hering, J. G. Looking Beyond Technology: An Integrated Approach to Water, Sanitation and Hygiene in Low Income Countries. *Environ. Sci. Technol.* **2014**, *48*, 9965-9970.

Research at private companies

A significant number of private companies operate in the area of water purification and wastewater treatment, with the goal to ensure safe drinking water to the residents of Switzerland. These companies include:

Pure Aqua[256] –

Although this company is based in the United States, it has strong operations in Switzerland, where it focuses on the development of reverse osmosis desalination and water treatment procedures.

Trunz Water Systems[257]

This company is a Switzerland-based family business that focuses on addressing water supply needs in remote areas, emergency response scenarios, defense ministries, and a variety of other settings. The company produces ultrafiltration units to purify fresh water sources; reverse osmosis units to desalinate brackish water; and highly energy efficient reverse osmosis systems to desalinate and purify seawater.

Filopur[258]

This company develops water filters that are marketed and sold directly to consumers, with the goal of enhancing the quality of their tap water and the desire of the customer to consume such tap water. The company also manufactures filters for showers, and touts the alleged benefits of using treated shower water for improved skin hydration and function.

Research at government agencies

Although the management of water supplies in Switzerland is highly decentralized, focused on the individual communities in which such supplies are located, there have been federal government efforts to conduct research in water purification and effectively managing the water supply.[259] Much of the role of the federal government has been to establish legally enforceable standards for drinking water purity, as well as standards for water used for other purposes. A variety of government documents on the state of water in Switzerland have also been reported.[260]

WATER RESEARCH IN THE UNITED KINGDOM

Introduction

Much of the work on water research within the United Kingdom is segregated by country, with parallel research efforts operating in England, Scotland, Wales, and Northern Ireland. Of note, the fact that the United Kingdom is surrounded by seawater provides an excellent opportunity to process that seawater via desalination and dramatically enhance the available water supply. Such desalination efforts have been operating in all four countries, with varying degrees of success. In general, the efforts towards obtaining clean drinking water through water purification and desalination efforts include research at academic institutions, private companies, and government agencies, all of which are discussed in detail below.

Research at academic institutions

Cardiff University

This is a public research university, located in Cardiff, Wales. Within the university, there is a highly active Water Research Institute, which conducts research in the area of water sustainability, water resource management during a period of climate change, and obtaining resilient freshwater sources for the improvement of community health and well-being.[261] The Institute is directed by Dr. Christopher Hubbard,[262] who has conducted extensive interdisciplinary work in the area of water research. Representative publications of Dr. Hubbard are listed below:

1. Wu, Y.; Cheng, Y.; Hubbard, C. G.; Hubbard, S.; Ajo-Franklin, J. B. Biogenic Sulfide Control by Nitrate and (Per)chlorate - A Monitoring and Modeling Investigation. *Chem. Geology* **2018**, *476*, 180-190.

2. Cheng, Y.; Hubbard, C. G.; Li, L.; Bouskill, N.; Molins, S.; Zheng, L.; Sonnenthal, E.; Conrad, M. E.; Engelbrektson, A.; Coates, J. D.; Ajo-Franklin, J. B. Reactive Transport Model of Sulfur Cycling as Impacted by Perchlorate and Nitrate Treatments. *Environ. Sci. Technol.* **2016**, *50*, 7010-7018.

Cranfield University[263]

– This is a research-based public university, located in Cranfield, England. As part of the university, there is a highly active Water Science Institute that focuses on water-related research.[264] The extensive research portfolio includes research in the development of a highly efficient nanomembrane toilet;[265] research to develop and restore rivers effectively;[266] and research in developing sustainable water resources in other, sometimes remote geographic locations.[267] The Institute is led by Professor Paul Jeffrey,[268] whose expertise is in the area of sustainable water use and re-use. Representative publications authored by Professor Jeffrey are listed below:

1. Goodwin, D.; Raffin, M.; Jeffrey, P.; Smith, H. M. Stakeholder Evaluations of Risk Interventions for Non-Potable Recycled Water Schemes: A Case Study. *Sci. Total Environ.* **2019**, *674*, 439-450.
2. Goodwin, D.; Raffin, M.; Jeffrey, P.; Smith, H. M. Informing Public Attitudes to Non-Potable Water Reuse - The Impact of Message Framing. *Water Res.* **2018**, *145*, 125-135.

University of Bath[269] - This public research university is located in Bath, Somerset, United Kingdom, and has a highly active Water Innovation and Research Center (WIRC).[270] Themes of the water reserch being conducted at the university include research in urban water management (including drinking water purification and wastewater treatment and management) as well as research in the area of water awareness and human behavior. The Institute is led by Director Professor Jan Hofman and Deputy Director Dr. Tom Arnot, both of whom have published extensively in the area of water research and water treatment procedures. Representative publications from each researcher are included below.

Representative publications from Professor Jan Hofman:

1. Hofman-Caris, R.; ter Laak, T.; Huiting, H.; Tolkamp, H.; de Man, A.; van Diepenbeek, P.; Hofman, J. Origin, Fate and Control of Pharmaceuticals in the Urban Water Cycle: A Case Study. *Water* **2019**, *11*, 1034.
2. Hofman-Caris, R.; Bertelkamp, C.; de Waal, L.; van den Brand, T.; Hofman, J.; Van Der Aa, R.; Van Der Hoek, J. P. Rainwater Harvesting for Drinking Water Production: A Sustainable and Cost-Effective Solution in the Netherlands? *Water* **2019**, *11*, 511.

Representative publications from Dr. Tom Arnot:

1. Proctor, K.; Petrie, B.; Barden, R.; Arnot, T.; Kasprzyk-Hordern, B. Multi-Residue Ultra-Performance Liquid Chromatography Coupled with Tandem Mass Spectrometry Method for Comprehensive Multi-Class Anthropogenic Compounds of Emerging Concern Analysis in a Catchment-Based Exposure-Driven Study. *Anal. Bioanal. Chem.* **2019**, *411*, 7061-7086.
2. Skouteris, G.; Arnot, T. C.; Jraou, M.; Feki, F.; Sayadi, S. Modeling Energy Consumption in Membrane Bioreactors for Wastewater Treatment in North Africa. *Water Environ. Res.* **2014**, *86*, 232-244.

Research at private companies

Desalination and commercial water filtration efforts in the United Kingdom have been executed primarily by private entities. A summary of those efforts is included below:

Thames Water Desalination Plant[271]

– This water desalination plant, located in Beckton, London, officially opened in June 2010. It was constructed via consortium of three private companies: Interserve (a United Kingdom based collection of companies),[272] Atkins Water (a global consulting firm),[273] and Acciona Agua (a Spanish-based company).[274] Criticism of the desalination plant included questions about sustainability as well as the potential for negative financial projections and high rates of leakage from the desalination plant.[275]

Pure Aqua[276] – This United States based company operates in a wide variety of countries world-wide, including in Switzerland (*vide supra*) as well as in the United Kingdom, where it produces highly effective water filters and reverse osmosis desalination products.

Lubron Water Technologies[277] - This UK-based company has branches in a variety of other countries, including the Netherlands, Belgium, and Germany, and focuses on providing both industrial and commercial water treatment services.

Osmio Water Company[278] - This company, located in Cliff, Kent, UK, focuses on developing water filters and on marketing and selling those filters directly to consumers, with the goal of improving the consumer's water quality and purity of the water supply.

PATENT REVIEW RESULTS

UNITED STATES INSTITUTION PATENTS

The number of large companies with patents in this area is extensive, and some of the notable players are listed below: BASF, BP, Chevron, Dow Global Technologies LLC, Eastman Kodak Company, Evoqua Water Technologies LLC, General Electric, Fujifilm Corp, Gradiant Corp, Halliburton Energy Services Inc., Hitachi LTD, Lockheed Corporation, Mitsubishi, Murtech Inc., Panasonic, Samsung, Shell Oil Company, Siemens Industry, Solvay, Tesla, and Toshiba.

Patents issued to smaller companies, universities, and government offices/laboratories are reviewed here by state:

Lawrence Livermore National Laboratory, located in Livermore, California, has obtained patents in the research area of capacitive charging and discharging related to osmosis. The titles of the three patents in this research area are listed below:

1. Capacitive Deionization Charge Transfer From One Capacitor Simultaneously To Multiple Capacitors
2. Multi-fluid Renewable Geo-energy Systems And Methods
3. Phased Charging And Discharging In Capacitive Desalination

Palo Alto Research and Development Company (PARC; https://www.parc.com/) has multiple patents in this research area. This company is based in Palo Alto, CA, and includes active research in the area of seawater desalination and in novel filter production.

A faculty member at the University of California, Yoram Cohen, has received two patents in this research area. These are listed below:

1. Apparatus, System and Method For Integrated Filtration And Reverse Osmosis Desalination

2. Fouling and Scaling Resistant Nano-structured Reverse Osmosis Membranes

Finally, the University of California has two patents in the area of water desalination. The titles of these patents are listed below:

1. Photovoltaic Desalination System
2. Apparatus and Method for Nanoporous Inorganic Membranes and Films, Methods of Making and Usage Thereof

US Department of the Interior

Andrew Patrick Murphy, an inventor who operated through the Department of the Interior, received a patent on intellectual property related to the synthesis of chlorine-resistant polyamides and membranes derived from such polymers. The application of such membranes to desalination was also discussed.

Yale University has received multiple patents in this research area to Robert L McGinnis, including some which are co-owned by Yale engineering Professor Menachem Elimelech (https://seas.yale.edu/faculty-research/faculty-directory/menachem-elimelech). Dr. McGinnis has also been granted several patents as part of his company, Oasys Water Inc (http://oasyswater.com/), which focuses on the development of novel osmosis processes. Additional patents to Yale University include patents in the area of improved osmosis systems.

University of Florida

University of Florida has a patent issued to James Klausner, a professor in the College of Engineering (https://vivo.ufl.edu/display/n12979), Renwei Mei (also a faculty member at the University of Florida: https://mae.ufl.edu/people/profiles/renwei-mei/), Srinivas Garimella, and Donald Ziegler. This patent is co-owned by Alcoa Inc. (https://www.alcoa.com/global/en/home.asp), a company based in Pennsylvania that operates primarily in the aluminum industry.

Desalitech

Professor Avi Efraty (who was previously a faculty member at University of Waterloo, University of Georgia, University of Rutgers and the Weizmann Institute) is the founder of Desalitech (www.desalitech.com), based in Massachusetts, and currently works as the Chief Technology Officer. He has a granted patent in the area of reverse osmosis.

MIT

MIT has **at least 20 patents** in this research area, including a patent issued to Rohit Karnik and Jongho Lee, entitled, "Liquid Filtration Using Pressure Difference Across A Hydrophobic Membrane." Of note, several of the patents have been jointly issued to the King Fahd University of Petroleum & Minerals, located in Saudi Arabia, highlighting the success of collaborative research efforts between the two institutions.

7AC Technologies

7AC Technologies (https://7actech.com/contact/) is an HVAC company located in Beverly, MA. They hold patents in a number of areas that are related to water purification and desalination technologies, including in the development of novel desiccants, and in the area of improved chilling procedures using water purification technologies. Listed co-inventors on their patents include Peter Vandermeulen, Jack Hanoka, and Chaim Hanoka.

Ionic Solutions

Ionic Solutions, LTD (http://www.ionicsolutions.com/contact.htm), is a small company based in Missouri that focuses on the area of water purification. They hold five patents in this research area, all of which are jointly held by co-inventor Azaroghly Yazdanbod.

Lytesyde LLC

Lytesyde, LLC was a company based in Nevada that has since dissolved. The company holds four patents in the area of water purification and desalination, and has two listed co-inventors: Bruce E.

Nadeau, Junior, and Kelly P. Rock. Co-Inventor Kelly P. Rock has two other patents on which they are listed as a co-inventor, both of which are held jointly with Micronic Technologies Inc. (https://www.micronictechnologies.com/).

Sunlight Photonics

Sunlight Photonics, a company based in Edison, New Jersey, has as its primary mission the development of unmanned autonomous vehicles. They hold two patents in the area of water purification and desalination, in the area of solar desalination and hybrid solar desalination systems.

CUNY Research Foundation

The City University of New York has two patents in the area of solar-powered water purification systems. The patent is held jointly by the Research Foundation of the City of New York and by Sun Fresh Water, LLC (http://sunfreshwater.com/), a company that is based in Florida. The faculty members involved in this research are Professor Ali Sadegh, a Professor in the Department of Mechanical Engineering of the City University of New York (CUNY) (https://www.ccny.cuny.edu/profiles/ali-sadegh), and Professor Jorge Gonzalez, also from the Department of Mechanical Engineering of CUNY (https://www.ccny.cuny.edu/profiles/jorge-gonzalez).

Drexel University

Young Cho, a faculty member in the Engineering Department at Drexel University (https://drexel.edu/engineering/about/faculty-staff/C/cho-young/) has two patents in this area entitled, "Desalination System And Process."

Ocean Thermal Energy Corporation

Ocean Thermal Energy Corporation (OTEC; https://otecorporation.com/contact/) is a company based in Pennsylvania that works in the area of ocean water desalination. The Chief Science Advisor, Stephen K. Oney, has multiple patents in the area of deep ocean and deep-water desalination.

Clemson University

Clemson University in South Carolina holds a patent in this area entitled, "Materials and Methods for Reducing Biofouling in Water Treatment Membrane Systems." This patent was granted to engineering professor Dr. Ezra Cates in July of 2019 (https://blogs.clemson.edu/environmental-engineering-and-earth-science/2019/07/08/cates-receives-patent-to-improve-water-treatment/).

Mansour S. Bader, located in Texas, has 12 patents granted in the area of water desalination. I was unable to find a particular affiliation for this inventor, as all contact information for him lists a PO Box in College Station, Texas.

Enviro Water Minerals Company

Enviro Water Minerals Company Inc (https://envirowaterminerals.com/), a company located in El Paso, Texas, has received 10 patents in the area of desalination and water purification. They have also been in featured in news articles related to cleaning up oil spill sites and regenerating safe drinking water. For examples, see the articles in the links listed below:

1. https://www.mrt.com/businessinsider/oilreport/article/Desalination-process-could-yield-fresh-water-from-13308827.php
2. https://www.prnewswire.com/news-releases/el-paso-boosts-its-drinking-water-supply-thanks-to-enviro-water-minerals-and-veolias-new-partnership-300224109.html
3. https://www.bloombergquint.com/business/risky-trash-to-treasure-ventures-backed-by-muni-bonds-hit-snags

University of Texas

Two seawater desalination patents are held by the University of Texas, based on work done by Professor Richard Crooks in the Department of Chemistry at the University of Texas at Austin.

George Washington University

George Washington University holds two patents in this area, both of which are granted to Professor Stuart Licht from the Department of Chemistry (https://chemistry.columbian.gwu.edu/stuart-licht). The titles of the patents are shown below, although the direct relevance to water purification research remains unclear:

1. Process for Iron and Steel Production
2. Process for Synthesis of Calcium Oxide

University of Wisconsin Research Foundation

The University of Wisconsin Madison Research Foundation Inc has two patents issued to Professor Zhen He, a faculty member in Civil Engineering and Mechanics at the University of Wisconsin Madison. These patents are entitled, "Osmotic Bioelectrochemical Systems," and "Integrated Photo-bioelectrochemical Systems."

FOREIGN INSTITUTION PATENTS

Patents by foreign institutions are also extensive and are briefly reviewed here by region.

AUSTRALIAN PATENTS

University of Melbourne has two patents that have been issued to Dimitrov Kolev Spas, a Professor of Chemistry (https://kolev.chemistry.unimelb.edu.au/sample-page/professor-spas-d-kolev/), Robert Walter Cattrall (also in the chemistry department: https://kolev.chemistry.unimelb.edu.au/current-group-members/professor-robert-cattrall/), and Youngsoo Cho, a graduate student at the University of Melbourne. The titles of these patents are listed below:

1. Polymer Inclusion Membranes for Use in Processes for Treating Thiocyanate Containing Aqueous Solutions
2. Process for Treating Thiocyanate Containing Aqueous Solutions

EGYPTIAN PATENTS

The Egypt Nanotechnology Center has a number of patents in the area of water desalination. The titles of these patents are listed below:

1. Photovoltaic Thermal Hybrid Systems and Method of Operation Thereof
2. Desalination System and Method for Desalination
3. Desalination System and Method for Desalination

Doped, Passivated Graphene Nanomesh, Method of Making the Doped, Passivated Graphene Nanomesh, And Semiconductor Device Including the Doped, Passivated Graphene Nanomesh

ISRAELI PATENTS

Ben Gurion University has a granted patent entitled, "Method and Apparatus for Effecting High Recovery Desalination with Pressure Driven Membranes," which they have filed jointly with Mekorot Water Company, the national water utility company of Israel (http://www.mekorot.co.il/Eng/Pages/default.aspx). They have another patent, filed jointly with the University of Colorado Regents, entitled, "Analyzing Ultrasonic Signals Using A Dynamic Window for An Early Detection of Scaling in Water Processing Equipment." A third patent in this research area from the Ben Gurion Research Authority is entitled, "Anion Exchange Membranes, Methods of Preparation and Uses."

Weizmann Institute of Science Commercialization Office, the Yeda Institute for Research and Development (https://www.yedarnd.com/), has a patent entitled, "Methods And Apparatuses For Decreasing The Co2 Concentration Of A Fluid."

KOREAN PATENTS

Significant research in Korea in this area is concentrated at the Korea Institute for Science and Technology, the Korean Advanced Institute for Science and Technology, the Korea Electric Power Company, and the Korea Institute for Machines and Materials. There are six patents granted to these institutions in the area of water desalination and purification, focused on new membrane development, water treatment, and advanced desalination protocols.

KUWAITI PATENTS

Research in Kuwait in the area of water desalination is highly advanced, mostly driven by a need to obtain pure drinking water for residents of this country. Patents to Kuwait-based researchers include several patents to the Kuwait Institute for Scientific Research as well as to Kuwait

University, with research areas focused on the conversion of seawater into drinking water through various desalination and osmosis processes.

QATARI PATENTS

Qatar, like many of the countries in the Middle East, relies on desalination to provide access to drinking water to the residents of the country. Research in this area has resulted in patents to Qatar University (http://www.qu.edu.qa/), and to the Qatar Foundation for Education, Science, and Community Development (https://www.qatardebate.org/en/page/about/about-qatar-foundation).

SAUDI ARABIAN PATENTS

The need for desalination and water purification in Saudi Arabia is significant in order to provide drinking water for the country's population. As such, it is not surprising that there is so much active research in the area. The main player for patents in this research area in Saudi Arabia is the King Fahd University of Petroleum & Minerals has received more than 20 patents in this research area. The patents include research areas such as improved desalination cell development, improved membrane development, novel methods for the removal of different classes of pollutants, and the use of solar power for water purification and recovery of unused water.

Other universities in Saudi Arabia also hold patents in this research area, including the University of King Saud (https://ksu.edu.sa/en/), with eight patents for the development of novel desalination systems, King Abdul Aziz University has a patent in the area of water desalination entitled, "Hydrogen-powered Desalination Plant." More significantly, the Saudi Arabia Oil Company (www.saudiaramco.com) holds a number of patents in this area, related to the fractionation of natural gas liquid and how such fractionation can provide power for distillations, cooling systems, and other water purification-related efforts.

TABLES OF RELEVANT GRANTS AND PATENTS FROM USASPENDING.GOV

Grants

Table 1. The top 20 (by amount of money) of grants in the United States, generated from the website usaspending.gov using the search term "water purification," for grants only, between the years of 2014 and 2019:

List number	Grant Title	Institution	State	Amount	Dates	Funding Agency
1	Low Temperature Microplasma UV Lighting Tiels for Water pUrification and Sterilization	Eden Park Illumination, Inc	IL	$2,149,212	2012-2017	Department of Energy
2	Novel Membranes and Systems for Industrial and Municipal Water Purification and Reuse	GE Global Research	NY	$2,036,205	2014-2018	Department of Energy
3	Supercritical Treatment Technology for	University of North Dakota	ND	$1,994,637	2018-2019	Department of Energy

	Water Purification					
4	Ultra-Low Energy, Low Cost Industrial Nanomembrane Manufacturing for Desalination, Water Purification, and Remediation	Covalent LLC	NV	$1,131,117	2015-2018	Department of Energy
5	Enhanced Non-Fouling Membranes for Water Purification and Recycling	Lynntech Inc	TX	$805,356	2014-2019	National Institutes of Health (NIEHS)
6	SBIR Phase II: Regenerable Adsorbent Filter for Water Purification	Amriton LLC	PA	$750,000	2017-2020	National Science Foundation, Division of Industrial Innovation

7	Scalable, Self-Powered Purification Technology for Brackish and Heavy Metal-Contaminated Water	Massachusetts Institute of Technology	MA	$749,119	2013-2014	Department of Energy
8	Enabling technologies for water reclamation in future long-term space missions: wastewater resource recovery for energy generation	University of Puerto Rico	PR	$720,000	2014-2018	National Aeronautics and Space Administration (NASA)
9	An Innovative Zero-Liquid Discharge Intermediate-Cold-Liquid Eutectic-Freeze	University of Oklahoma	OK	$608,333	2019-2022	Department of Energy (ARPA-E)

	Desalination System					
10	Targeted Lead (Pb) Removal for Drinking Water Purification Using Incion	Powertech Water, Inc.	KY	$547,468	2017-2021	National Institutes of Health (NIEHS)
11	Carboxycellulose Nanofibers from Underutilized Biomasses for Water Purification	Research Foundation for the State University of New York	NY	$540,000	2018-2022	National Science Foundation
12	Biological Diversity and Ecological Forecasting	University System of Maryland	MD	$425,728	2017-2020	National Aeronautics and Space Administration (NASA)
13	Improving IOP measurement uncertainties for PACE ocean color remote sensing applications.	Florida Atlantic University	FL	$386,143	2015-2019	National Aeronautics and Space Administration (NASA)

14	SUSCHEM: Novel Biomimetic Materials for Water Purification: Perchlorate Treatment	University of Illinois	IL	$385,612	2013-2018	National Science Foundation
15	Functional Porous Organic Polymers as Advanced Decontamination Materials for Water Purification	University of South Florida	FL	$329,947	2017-2020	National Science Foundation
16	GOALI: Application of an Innovative Anti-Biofilm Technology for Overcoming Biofouling on Water	University of North Carolina at Chapel Hill	NC	$325,975	2013-2018	National Science Foundation

	Purification Membranes					
17	Hyperion Water Demonstration Facility	City of Los Angeles	CA	$300,000	2016-2017	Bureau of Reclamation
18	High Stable Nanoparticle-Doped Metal-Organic Frameworks for Applications in Water Purification	Virginia Union University	VA	$286,319	2019-2022	National Science Foundation
19	Construction of a Biological Treatment System at the Joint Water Purification Plant	Cottonwood Water and Sanitation District	CO	$282,385	2016-2019	Bureau of Reclamation
20	Land-Cover/ Land-Use Program	University of Puerto Rico	PR	$268,829	2013-2017	National Aeronautics and Space Administration (NASA)

Table 2. The top 20 (by amount of money) of grants in the United States, generated from the website usaspending.gov using the search term "desalination," for grants only, between the years of 2014 and 2019:

List number	Grant Title	Institution	State	Amount	Dates	Funding Agency
1	SBIR: A Very Low Cost Process for the Manufacture of Ti Heat Exchanger CO	Materials & Electrochemical Research Corp.	AZ	$2,018,090	2010-2013	Department of Energy
2	Zero Liquid Discharge Water Desalination Process Using Humidification-Dehumidification in a Thermally-Actuated Transport Reactor	Oregon State University	OR	$2,000,000	2018-2020	Department of Energy
3	Full Scale Testing, Characterization, System Optimization & Demonstration of	Texas Tech University	TX	$1,992,446	2006-2013	Department of Energy

	Grid Connected Wind Turbines & Wind Powered Desalination					
4	Hawaii Solar Desalination Project	Business, Economic Development and Tourism	HI	$1,928,238	2018-2019	Department of Energy
5	Energy Where It Matters: Delivering Heat to the Membrane/Water Interface for Enhanced Thermal Desalination	University of California Los Angeles	CA	$1,709,744	2018-2020	Department of Energy
6	Low Cost Desalination Using Nanophotonics Enhanced Direct Solar Membrane Distillation	William Marsh Rice University	TX	$1,699,988	2018-2020	Department of Energy
7	INFEWS/T2: Improving Crop Yield and Soil Salinity by Cost-	University of North Texas	TX	$1,575,380	2019-2024	National Science Foundation

	Effective Integration of Microbial Community, Hydrology, Desalination, and Renewable Power					
8	TAS::89 0222::TAS; New, Development of Ultrafiltration Membrane-Separation-Technology for Energy-Efficient Water Treatment and Desalination Process	Nevada System of Higher Education	NV	$1,554,164	2010-2016	Department of Energy
9	Middle East Desalination Research Center Operations	Middle East Desalination Research Center		$1,500,000	2017-2021	Department of State
10	Skytrough Vacuum Membrane: An Extreme Low-Cost Solar Thermal	Skyfuel, Inc.	CO	$1,468,984	2018-2019	Department of Energy

	Collector for Desalination					
11	Molecularly Precise Nanoporous Desalination Membranes	TDA Research, Inc.	CO	$1,150,000	2016-2020	Department of Energy
12	Nanomembrane Interactive Forward Osmosis (FO) Polymers for Desalination and Remediation	Covalent LLC	NV	$1,149,976	2017-2020	Department of Energy
13	Solar Thermal Assisted Vacuum Freezing Desalination of Seawater at the Triple Point	Advanced Cooling Technologies, Inc.	PA	$1,149,954	2016-2019	Department of Energy
14	Membraneless Water Desalination System	Reactive Innovations, LLC	MA	$1,149,886	2014-2017	Department of Energy
15	Wave-Powered Desalination System	Resolution Marine Energy, Inc.	MA	$1,147,416	2017-2020	Department of Energy

16	Ultrahigh-Efficiency Capacitive Devices for Continuous Water	Mainstream Engineering Corporation	FL	$1,132,558	2014-208	Department of Energy
17	Feasibility Studies, Long Beach Desalination Research and Development Project	City of Long Beach	CA	$1,132,108	2002-2022	Department of the Interior
18	Ultra-Low Energy, Low Cost Industrial Nanomembrane Manufacturing for Desalination, Water Purification, and Remediation	Covalent LLC	NV	$1,131,117	2015-2018	Department of Energy
19	PEM Based Vacuum Desalination System	Xergy Inc.	NJ	$1,083,709	2016-2019	Department of Energy
20	The Internal Compound Parabolic Concentrator (ICPC): A Novel	University of California Merced	CA	$1,081,793	2018-2020	Department of Energy

	Low Cost Solar Thermal Collection System for Desalination Processes					

Contracts

Table 3. The top 20 (by amount of money) of contracts in the United States, generated from the website usaspending.gov using the search term "water purification," for contracts only, between the years of 2014 and 2019:

List number	Contract Title	Recipient	State	Amount	Dates	Funding Agency
1	Detail, Design, and Construction of LPD 17 Class Ship (LPD 22)	Huntington Ingalls Incorporated	MS	$10,030,868,842.71	2006 - 2022	Department of Defense (Department of the Navy)
2	F07 Execution Planning for USS Boxer (LHD 4) FY 07 Non Dry-Docking-CNO Scheduled Availability.	National Steel and Shipbuilding Company	CA	$892,382,358.86	2007 - 2019	Department of Defense (Department of the Navy)

3	EGI IGF::CL::IGF INP II To Obligate the Amount of $17,335,629 To Issue the Middle Area Gaza Desalination Plant Expansion Project Under the INP II Program	Blumont Engineering Solutions, Inc.	WI	$17,335,629.00	2016 - 2019	Agency for International Development
4	IGF::OT::IGF South Ferris Water Supply Desalination Project Brackish Water Wells No. 95 and No. 96 Eastern Municipal Water District - Riverside County, California	Hemet Manufacturing Company, Inc.	CA	$13,111,677.27	2015 - 2019	Department of Defense (Department of the Army)
5	IGF::OT::IGF	McKean Defense Group, LLC	PA	$3,418,693.00	2017 - 2020	Department of Defense (Departmen

						t of the Navy)
6	IGF::OT::IGF Low Power Water Purification and Desalination System. Small Business Innovation Research (SBIR) Phase II	Lynntech Inc	TX	$1,498,563.82	2016 - 2019	Department of the Interior
7	TAS::96 3134::TAS ARRA::YES::ARR A Recovery Project#174734::R P# Cambria Desalination Facility, San Luis Obispo County, California	CDM Federal Programs Corporation	MO	$1,153,663.31	2010 - 2014	Department of Defense (Departmen t of the Army)
8	ER-2216	Pennsylvania State University, The	PA	$1,010,616.00	2012 - 2017	Department of Defense (Departmen

						t of the Army)
9	Development of Desalination Membranes – Base and Four Option Years	Separation Systems Technology	CA	$660,419.63	2010 - 2014	Department of the Interior
10	IGF::OT::IGF 270' WMEC Reverse Osmosis Prototype, Fabrication, and Installation	Aqua-Chem, Inc	TN	$415,889.81	2017 - 2018	Department of Homeland Security (US Coast Guard)
11	Second Option Period for the DB IDIQ IGF::OT::IGF DB CON Desalination Plant ST. Vincents, the Grenadines	Strategic Consulting Alliances, LLC	MD	$388,841.53	2016 - 2017	Department of Defense (Department of the Navy)
12	IGF::OT::IGF CGC Spencer Reverse Osmosis Installation	Aqua-Chem, Inc	TN	$350,354.61	2018	Department of Homeland Security

						(US Coast Guard)
13	IGF::OT::IGF CGC Campbell Reverse Osmosis Installation FY18	Aqua-Chem, Inc	TN	$350,354.61	2018	Department of Homeland Security (US Coast Guard)
14	Vessel, Desalination	Watermaker Store, Inc	CA	$292,430.60	2016 - 2017	Department of Defense (Defense Logistics Agency)
15	Broad Agency Announcement Research and Development Contract for Process Analysis of Energy Recovery Approaches for Small-Scale Expeditionary	University of California, Los Angeles	CA	$208,417.25	2011 - 2015	Department of Defense (Department of the Navy)

	Desalination Systems					
16	IGF::OT::IGF Future Naval Capabilities Advanced Shipboard Desalination Test Support	JSL Technologies Inc	CA	$199,916.66	2015 - 2016	Department of Defense (Department of the Navy)
17	Purchase of Stock Parts for Reverse Osmosis Desalination Units for Medium Endurance Cutter Product Line (MECPL)	Parker-Hannifin Corporation	CA	$166,346.99	2019	Department of Homeland Security (US Coast Guard)
18	IGF::OT::IGF Other Functions - DARPA SBIR 15.1 – Low Power Water Purification and Desalination System	Lynntech, Inc.	TX	$150,000.00	2015 - 2016	

19	Vessel, Desalination	Watermaker Store, Inc.	CA	$146,215.30	2016	
20	Vessel, Desalination	Watermaker Store, Inc.	CA	$146,215.30	2016	

TABLES OF RELEVANT PATENTS FROM LENS.ORG

Table 4. Top Inventors in the area of "water purification" from 2013-2020, measured by the number of US granted patents issued to that inventor in that time

Inventor name	Number of patents	Relevant web links*
Austin L. Gurney	664	https://www.linkedin.com/in/austin-gurney-2b8097176/
William I. Wood	646	https://www.linkedin.com/in/william-wood-30b019114/
Audrey Goddard	620	https://www.linkedin.com/in/audrey-goddard-176a914/
Paul Godowski	580	https://www.linkedin.com/in/paul-godowski-0b1b296/
Jennifer Hillman	320	
Satoshi Seo	313	https://www.researchgate.net/lab/Satoshi-Seo-Lab
Victoria Smith	299	https://www.linkedin.com/in/victoria-smith-6460556b/
David M. Goldenberg	294	https://www.researchgate.net/scientific-contributions/38337014_David_M_Goldenberg
Craig A. Rosen	294	https://www.alignable.com/lake-placid-fl/a-all-hour-cleanup-co-inc
Neil C. Corley	284	https://patents.justia.com/inventor/neil-c-corley

*These are my best estimates of the correct links matched to the correct people, but they are not 100% vetted for accuracy.

Table 5. Top Inventors in the area of "desalination" from 2013-2020, measured by the number of US granted patents issued to that inventor in that time

Inventor name	Number of patents	Relevant web links*
Prakash Narayan Govindan	41	https://www.linkedin.com/in/prakash-govindan-87168310/?originalSubdomain=sg
Bart Lipkens	38	https://www.linkedin.com/in/bart-lipkens-7427967/
Avetik Markosyan	35	https://www.researchgate.net/profile/Avetik_Markosyan
Maximus G St. John	33	https://patents.justia.com/inventor/maximus-g-st-john?page=2
Steven Lam	30	https://www.linkedin.com/in/steven-lam-engineer/?originalSubdomain=sg
Chunqing Liu	28	https://patents.justia.com/inventor/chunqing-liu
Li-Shiang Liang	24	https://patents.justia.com/inventor/li-shiang-liang
Peter F Vandermeulen	24	http://www.patentsencyclopedia.com/inventor/peter-f-vandermeulen-newburyport-us-1/
John H Lienhard	22	https://meche.mit.edu/people/faculty/LIENHARD@MIT.EDU
Jason Dionne	21	https://patents.justia.com/assignee/flodesign-sonics-inc

*These are my best estimates of the correct links matched to the correct people, but they are not 100% vetted for accuracy.

WEB OF SCIENCE TABLES

Searched for highly cited papers in the field of "water purification"; search restricted to English language journal articles

Searched for highly cited papers in the field of "desalination"; search restricted to English language journal articles

REFERENCES

[1] https://www.brown.edu/research/projects/superfund/

[2] https://www.brown.edu/news/2017-08-10/water

[3] https://news.brown.edu/articles/2015/09/superfund

[4] http://news.mit.edu/2017/electrochemical-clear-pollutants-water-0510

[5] https://d-lab.mit.edu/research/water-filters

[6] http://energy.mit.edu/news/turning-desalination-waste-into-a-useful-resource/

[7] http://news.mit.edu/2016/solar-powered-desalination-clean-water-india-0718

[8] https://mitmrsec.mit.edu/

[9] http://energy.mit.edu/

[10] http://news.mit.edu/2019/water-innovation-prize-methane-wastewater-0423

[11] http://news.mit.edu/2018/new-system-recovers-fresh-water-power-plants-0608

[12] https://www.technologyreview.com/s/413046/water-splitting-company-founded/

[1313] https://sageguardsolutions.com/ http://meche.mit.edu/featured-classes/desalination-and-water-purification

[14] https://technology.nasa.gov/featurestory/safeguard-license

[15] https://www.nasa.gov/sites/default/files/atoms/files/benefits-for-humanity_third.pdf

[16] https://www.concern4kids.org/

[17] https://www.nasa.gov/centers/marshall/multimedia/photos/2006/photos06-080.html

[18] https://www.nasa.gov/sites/default/files/atoms/files/fy19_nasa_budget_estimates.pdf

[19] https://chemicaleng.njit.edu/faculty/sirkar

[20] https://chemicaleng.njit.edu/faculty/khusid

[21] (a) Puranik, A. A.; Rodrigues, L. N.; Chau, J.; Li, L.; Sirkar, K. K. Porous hydrophobic-hydrophilic composite membranes for direct contact membrane distillation. *J. Membrane Sci.* **2019**, *591*, 117225; (b) Chau, J.; Singh, D.; Sirkar, K. K. 110th Anniversary: Liquid Separation Membranes Based on Nanowire Substrates for Organic Solvent Nanofiltration and Membrane Distillation. *Industrial Engineering Chem. Res.* **2019**, *58*, 14350-14356; (c) Yao, N.; Khusid, B.; Sirkar, K. K.; Dehn, D. J. Nanoparticle filtration through microporous ECTFE membrane in an alcoholic solution. *Separation Purification Technol.* **2019**, *210*, 754-763.

[22] (a) Sirkar, K.; Singh, D.; Li, L.; McEvoy, T. J. Hollow fiber membrane module for direct contact membrane distillation-based desalination. *PCT Int. Appl.* **2018**, WO 2018195534 A1 20181025; (b) Sirkar, K.; Pfeffer, R.; Chen, D.; Singh, D. Porous hollow fiber antisolvent crystallization-based continuous method of polymer coating on submicron and nanoparticles. *U.S. Pat. Appl. Publ.* **2016**, US 20160166512 A1 20160616.

[23] https://math.njit.edu/faculty/lcumming

[24] https://www.nsf.gov/pubs/2016/nsf16099/nsf16099.jsp

[25] https://www.chemistry.northwestern.edu/people/core-faculty/profiles/william-dichtel.html

[26] https://cyclopure.com/

[27] https://www.waterworld.com/drinking-water/treatment/article/14040084/cyclopure-wins-1-million-phase-ii-sbir-grant

[28] https://cyclopure.com/product/water-test-kit-pro/

[29] https://www.scholars.northwestern.edu/en/projects/cbet-epsrc-a-game-changing-approach-for-tunable-membrane-developm-4

[30] bwc.berkeley.edu

[31] https://ce.berkeley.edu/people/faculty/sedlak

[32] https://erg.berkeley.edu/people/ray-isha/

[33] https://samueli.ucla.edu/people/yoram-cohen/

[34] (a) Lee, T.; Rahardianto, A.; Cohen, Y. Multi-cycle operation of semi-batch reverse osmosis (SBRO) desalination. *J. Membrane Sci.* **2019**, *588*, 117090; (b) Lee, T.; Rahardianto, A.; Cohen, Y. Flexible reverse osmosis (FLERO) desalination. *Desalination* **2019**, *452*, 123-131; (c) Cohen, Y.; Semiat, R.; Rahardianto, A. A perspective on reverse osmosis water desalination: Quest for sustainability. *AIChE J.* **2017**, *63*, 1771-1784.

[35] (a) Cohen, Y.; Christofides, P. D.; Gu, H.; Gao, L. X.; Rahardianto, A. Self-adaptive control and optimization of membrane filtration. *U.S. Pat. Appl. Publ.* **2017**, US 20170209834 A1 20170727.; (b) Cohen, Y.; Mccool, B. C.; Rahardianto, A. Method and system for high recovery water desalting. *U.S. Pat. Appl. Publ.* **2011**, US 20110155665 A1 20110630.

[36] https://www.ee.ucla.edu/subramanian-s-iyer/

[37] www.nrgtekusa.com/

[38] (a) Iyer, S. Hybrid systems and methods with forward osmosis and electrodeionization using high-conductivity membranes. *U.S. Pat. Appl. Publ.* **2017**, US 20170129796 A1 20170511; (b) Fuchigami, K.; Tsuji, T.; Tomura, K.; Fujiwara, S.; Kunugi, M.; Kato, A.; Iyer, S. Temperature-sensitive absorbent, water treatment method, and water treatment apparatus. *PCT Int. Appl.* **2015**, WO 2015156404 A1 20151015; (c) Iyer, S. Systems and methods for forward osmosis fluid purification. *U.S. Pat. Appl. Publ.* **2013**, US 20130032538 A1 20130207.

[39] https://cwe.unm.edu/

[40] http://www.unm.edu/~howe/

[41] http://climatecollege.unimelb.edu.au/recent-australian-droughts-may-be-worst-800-years

[42] http://www.awa.asn.au/AWA_MBRR/Publications/Fact_Sheets/Desalination_Fact_Sheet.aspx

[43] http://www.awa.asn.au/AWA_MBRR/Publications/Fact_Sheets/Desalination_Fact_Sheet.aspx

[44] https://www.arc.gov.au/grants/linkage-program/special-research-initiatives/pfas-remediation-research-program

[45] https://rms.arc.gov.au/RMS/Report/Download/Report/a3f6be6e-33f7-4fb5-98a6-7526aaa184cf/201

[46] https://www.adelaide.edu.au/environment/wrc/

[47] https://www.adelaide.edu.au/directory/justin.brookes

[48] https://sciences.adelaide.edu.au/biological-sciences/research/ecology-environmental-science

[49] (a) Brookes, J. D.; Tomlinson, A. Pick your poison: do cyanotoxins or disinfection by-products pose the greater risk? *Inland Waters* **2019**, *9*, 345-347; (b) Wu, T.; Qin, B.; Brookes, J. D.; Yan, W.; Ji, X.; Feng, J. Spatial Distribution of Sediment Nitrogen and Phosphorus in Lake Taihu from a Hydrodynamics-Induced Transport Perspective. *Sci. Total Environ.* **2019**, *650*, 1554-1565.

[50] https://www.ncl.ac.uk/sustainability/about/staff/profile/philipmcgowan.html#background

[51] https://research.ncl.ac.uk/bewise/

[52] https://www.ncl.ac.uk/engineering/staff/profile/russelldavenport.html#background

[53] (a) McGowan, P. J. K. Conservation Mapping the terrestrial human footprint. *Nature* **2016**, *537*, 172-173; (b) Geary, M.; Fielding, A. H.; McGowan, P. J. K.; Marsden, S. J. Scenario-led habitat modelling of land use change impacts on key species. *PLoS One* **2015**, *10*, e0142477/1-e0142477/17.

[54] (a) Komolafe, O.; Bowler, B.; Dolfing, J.; Mrozik, W.; Davenport, R. J. Quantification of polybrominated diphenyl ether (PBDE) congeners in wastewater by gas chromatography with electron capture detector (GC-ECD). *Anal. Methods* **2019**, *11*, 3474-3482; (b) Acharya, K.; Khanal, S.; Pantha, K.; Amatya, N.; Davenport, R. J.; Werner, D. A comparative assessment of conventional and molecular methods, including MinION nanopore sequencing, for surveying water quality. *Sci. Reports* **2019**, *9*, 1-11.

[55] https://www.unisa.edu.au/IT-Engineering-and-the-Environment/Natural-and-Built-Environments/Our-research/

[56] https://people.unisa.edu.au/Mizanur.Rahman

[57] Chow, C. W. K.; Liu, J.; Li, J.; Swain, N.; Reid, K.; Saint, C. P. Development of smart data analytics tools to support wastewater treatment plant operation. *Chemometrics Intelligent Lab. Systems* **2018**, *177*, 140-150.; (b) Kabir, Md. S.; Rahman, Md. M.; Ahmed, S. M. F. Physico-chemical and bacteriological attributes of wastewater in Savar, Dhaka, Bangladesh. *IOSR J. Environ. Sci. Toxicol. Food Technol.* **2015**, *9*, 06-10.

[58] https://sydney.edu.au/engineering/our-research/energy-resources-and-the-environment/water-resources-and-the-environment.html

[59] Luiz, A.; Spencer, E.; McClure, D. D.; Coster, H. G. L.; Barton, G. W.; Kavanagh, J. M. Membrane selection for the desalination of bio-refinery effluents using electrodialysis. *Desalination* **2018**, *428*, 1-11.

[60] Chen, Y.-M.; Hobson, P.; Burch, M. D.; Lin, T.-F. In situ measurement of odor compound production by benthic cyanobacteria. *J. Environ. Monitoring* **2010**, *12*, 769-775.

[61] Lim, S. Y.; Liang, Y. Y.; Fimbres Weihs, G. A.; Wiley, D. E.; Fletcher, D. F. A CFD study on the effect of membrane permeance on permeate flux enhancement generated by unsteady slip velocity. *J. Membrane Sci.* **2018**, *556*, 138-145.

[62] https://changingclimate.ca/CCCR2019/

[63] https://globalnews.ca/news/5948593/climate-change-canada-un-report/

[64] https://globalnews.ca/news/6190102/fresh-water-canada-climate-change/

[65] https://www.canada.ca/en/government/grants-funding.html

[66] https://cwra.org/en/

[67] https://carleton.ca/gwi/

[68] https://carleton.ca/cee/people/ormeci-banu/

[69] (a) Almomani, F. A.; Ormeci, B.; Kiely, P. Improving the performance of attached-growth wastewater treatment processes by altering the support media surface. *J. Environ. Engineering* **2019**, *145*, 4019045; (b) Diak, J.; Ormeci, B. Stabilisation and dewatering of primary sludge using ferrate(VI) pre-treatment followed by freeze-thaw in simulated drainage beds. *J. Environ. Management* **2018**, *216*, 406-420.

[70] (a) Ormeci, B.; Glover, A.; Kollu, K. Device for detecting algae concentration using first derivative of visible light absorbance for water pollution monitoring. *PCT Int. Appl.* **2017**, WO 2017201622 A1 20171130; (b) Abu-Orf, M.; Ormeci, B. Systems and methods for water and wastewater sludge treatment using floc or network strength. *U.S. Pat. Appl. Publ.* **2003**, US 20030230538 A1 20031218.

[71] https://newsroom.carleton.ca/story/water-research/

[72] https://www.mcgill.ca/h2oinnovation/

[73] https://www.mcgill.ca/iwrm/

[74] https://water.usask.ca/

[75] https://jayfamiglietti.com/

[76] https://water.utoronto.ca/about/leadership/

[77] (a) Martin, J. T.; Kolliopoulos, G.; Papangelakis, V. G. An improved model for membrane characterization in forward osmosis. *J. Membrane Sci.* **2019**, *Ahead of Print*; DOI: 10.1016/j.memsci.2019.117668; (b) Kolliopoulos, G.; Martin, J. T.; Papangelakis, V. G. Energy requirements in the separation-regeneration step in forward osmosis using TMA-CO2-H2O as the draw solution. *Chem. Engineering Res. Design* **2018**, *140*, 166-174.

[78] https://uwaterloo.ca/water-institute/

[79] https://uwaterloo.ca/water-institute/about/people/rbrouwer

[80] https://uwaterloo.ca/global-water-futures/

[81] Zhongfan, Z.; Dingzhi, P.; Hongrui, W. Seawater desalination in China: an overview. J. Water Reuse Desalination 2019, 9, 115-132.

[82] https://www.technologyreview.com/s/601861/chinas-massive-effort-to-purify-seawater-is-drying-up/

[83] https://www.researchgate.net/lab/Beijing-Key-Lab-of-Urban-Hydrological-Cycle-and-Sponge-City-Technology-Zongxue-Xu

[84] https://www.fudan.edu.cn/en/

[85] http://environment.fudan.edu.cn/en/Show.aspx?info_lb=265&flag=225&info_id=271

[86] http://environment.fudan.edu.cn/en/Show.aspx?info_lb=265&flag=225&info_id=280

[87] https://www.journals.elsevier.com/water-research/editorial-board/professor-jun-ma

[88] https://www.researchgate.net/profile/Nan-Qi_Ren

[89] https://www.researchgate.net/profile/Cm_Kao/research

[90] https://www.researchgate.net/scientific-contributions/2046490308_Fangang_Meng

[91] https://www.tsinghua.edu.cn/publish/thu2018en/newthuen_cnt/03-research-1.html

[92] https://biography.omicsonline.org/china/tsinghua-university/jianlong--wang-1559045

[93] https://www.tsinghua.edu.cn/publish/enven/6287/index.html

[94] https://www.researchgate.net/profile/Han-Qing_Yu

[95] http://dsxt.ustc.edu.cn/zj_ywjs.asp?zzid=385

[96] https://en.scms.ustc.edu.cn/2016/0316/c7149a89560/page.htm

[97] https://en.whu.edu.cn/Schools1.htm

[98] https://www.researchgate.net/profile/Hui_Zhang294

[99] https://www.researchgate.net/profile/Feng_Wu5/research

[100] http://sres.whu.edu.cn/info/1089/4916.htm

[101] https://www.suez.com/en

[102] https://www.suez.com/en/our-offering/success-stories/our-references/barcelone-seawater-desalination-plant

[103] http://www.sidem-desalination.com/about-us/veolia-group/

[104] https://www.aquatechtrade.com/news/desalination/worlds-largest-desalination-plants/

[105] https://www.psl.eu/en/psl-environment

[106] Hubert, N.; Espiau, B.; Meyer, C.; Planes, S. Identifying the ichthyoplankton of a coral reef using DNA barcodes. *Molec. Ecology Resources* **2015**, *15*, 57-67.

[107] http://www.cnrs.fr/en/artificial-bio-inspired-membranes-water-filtration

[108] Kocsis, I.; Sorci, M.; Vanselous, H.; Murail, S.; Sanders, S. E.; Licsandru, E.; Legrand, Y.-M.; van der Lee, A.; Baaden, M.; Petersen, P. B.; et al Oriented chiral water wires in artificial transmembrane channels. *Sci. Adv.* **2018**, *4*, eaao5603/1-eaao5603/9.

[109] http://documentation.sorbonne-universites.fr/en/about-us.html

[110] Manes, C.-L. de O.; Khan, M. T.; Molina, V. G.; Croue, J.-P. The use of microbial and chemical analyses to characterize the variations in fouling profile of seawater reverse osmosis (SWRO) membrane. *Desalination Water Treatment* **2013**, *51*, 944-949.

[111] https://www.utc.fr/~pauss/research.php

[112] Bai, H.; Cochet, N.; Pauss, A.; Lamy, E. DLVO, hydrophobic, capillary and hydrodynamic forces acting on bacteria at solid-air-water interfaces: Their relative impact on bacteria deposition mechanisms in unsaturated porous media. *Colloids Surfaces B* **2017**, *150*, 41-49.

[113] https://www.ladhyx.polytechnique.fr/en/publications/revues/2019/

[114] https://www.universite-paris-saclay.fr/laboratoires/centre-international-de-recherche-sur-lenvironnement-et-le-developpement

[115] Hossard, L.; Guichard, L.; Pelosi, C.; Makowski, D. Lack of evidence for a decrease in synthetic pesticide use on the main arable crops in France. *Sci. Total Environ.* **2017**, *575*, 152-161.

[116] Lammoglia, S.-K.; Makowski, D.; Moeys, J.; Justes, E.; Barriuso, E.; Mamy, L. Sensitivity analysis of the STICS-MACRO model to identify cropping practices reducing pesticides losses. *Sci. Total Environ.* **2017**, *580*, 117-129.

[117] https://www.businesswire.com/news/home/20120713005327/en/Research-Markets-Germany-Desalination-Plants-Market-Forecast

[118] https://www.bestwater-germany.com/

[119] https://www.techsciresearch.com/news/3192-bestwater-ranked-1-in-germany-water-purifiers-market.html

[120] https://www.hamburg.com/residents/housing/12792124/water-quality/

[121] https://www.seccua.com/about-seccua/

[122] https://www.hs-karlsruhe.de/en/the-university/

[123] https://www.hs-karlsruhe.de/io/international-cooperation/water-treatment-cooperation/

[124] https://www.hs-karlsruhe.de/en/research/hska-research-institutions/institute-of-refrigeration-air-conditioning-and-environmental-engineering/

[125] Klinger, J.; Sacher, F.; Brauch, H. J.; Maier, D. Determination of organic phosphonates in aqueous samples using liquid chromatography/particle-beam mass spectrometry. *Acta Hydrochimica et Hydrobiologica* **1997**, *25*, 79-86.

[126] Sacher, F.; Raue, B.; Klinger, J.; Brauch, H.-J. Simultaneous determination of Cr(III) and Cr(VI) in ground- and drinking waters by IC-ICP-MS. *Int. J. Environ. Analytical Chem.* **1999**, *74*, 191-201.

[127] https://www.campus-elgouna.tu-berlin.de/water/v_menu/department_of_water_engineering/

[128] Zouboulis, A. I.; Tzoupanos, N. Alternative cost-effective preparation method of polyaluminum chloride (PAC) coagulant agent: Characterization and comparative application for water/wastewater treatment. *Desalination* **2010**, *250*, 339-344.

[129] Deliyanni, E.; Arabatzidou, A.; Tzoupanos, N.; Matis, K. A. Adsorption of Pb2+ using mesoporous activated carbon and its effects on surface modifications. *Adsorption Sci. Technol.* **2012**, *30*, 627-645.

[130] https://wasserchemie.ebi.kit.edu/english/286.php

[131] https://wasserchemie.ebi.kit.edu/english/439_2704.php

[132] Sturm, M. T.; Herbort, A. F.; Horn, H.; Schuhen, K. Comparative study of the influence of linear and branched alkyltrichlorosilanes on the removal efficiency of polyethylene and polypropylene-based microplastic particles from water. *Environ. Sci. Pollution Res.* **2020**, *Ahead of Print*; DOI: 10.1007/s11356-020-07712-9

[133] Jeihanipour, A.; Shen, J.; Abbt-Braun, G.; Huber, S. A.; Mkongo, G.; Schaefer, A. I. Seasonal variation of organic matter characteristics and fluoride concentration in the Maji ya Chai River (Tanzania): Impact on treatability by nanofiltration/reverse osmosis. *Sci. Total Environ.* **2018**, *637-638*, 1209-1220.

[134] https://en.wikipedia.org/wiki/Minjur_Seawater_Desalination_Plant

[135] https://www.pv-magazine.com/2019/04/08/india-takes-to-solar-powered-desalination/

[136] Singh, A. K.; Poonia, S.; Jain, D.; Mishra, D. Performance evaluation and economic analysis of solar desalination device made of building materials for hot arid climate of India. *Desalination Water Treatment* **2019**, *141*, 36-41.

[137] Shrivastava, Brajesh K. Technological innovation in the area of drinking water for treatment of saline water. *Asian J. Water, Environ. Pollution* **2016**, *13*, 37-44.

[138] https://en.wikipedia.org/wiki/Chennai_MetroWater_Supply_and_Sewage_Board

[139] https://www.thehindubusinessline.com/info-tech/xiaomi-launches-a-made-for-india-smart-water-purifier/article29440683.ece

[140] https://www.livpure.com/blog/has-water-purifier-become-an-absolute-necessity-in-india-may-be-yes

[141] https://www.kent.co.in/blog/3-facts-about-drinking-water-in-india-why-a-water-purifier-is-a-must/

[142] http://www.idfc.com/pdf/report/2011/Chp-19-Water-Pollution-in-India-An-Economic-Appraisal.pdf

[143] Sprenger, C.; Lorenzen, G.; Grunert, A.; Ronghang, M.; Dizer, H.; Selinka, H.-C.; Girones, R.; Lopez-Pila, J. M.; Mittal, A. K.; Szewzyk, R. Removal of indigenous coliphages and enteric viruses during riverbank filtration from highly polluted river water in Delhi (India). *J. Water Health* **2014**, *12*, 332-342.

[144] https://www.iisc.ac.in/

[145] http://www.icwar.iisc.ac.in/

[146] http://civil.iisc.ernet.in/~pradeep/

[147] Rehana, S.; Mujumdar, P. P. An imprecise fuzzy risk approach for water quality management of a river system. *J. Environ. Management* **2009**, *90*, 3653-3664.

[148] Rehana, S.; Mujumdar, P. P. Climate change induced risk in water quality control problems. *J. Hydrology* **2012**, *444-445*, 63-77.

[149] Mondal, A.; Mujumdar, P. P. Regional hydrological impacts of climate change: implications for water management in India. *Proc. Int. Association Hydrological Sci.* **2015**, *366*, 34-43.

[150] https://www.thehindu.com/news/cities/bangalore/namma-bengaluru-resources-being-pooled-to-predict-rain-and-tackle-floods/article29183951.ece

[151] https://economictimes.indiatimes.com/news/politics-and-nation/the-precarious-situation-of-indias-water-problem/articleshow/57965416.cms?utm_source=contentofinterest&utm_medium=text&utm_campaign=cppst

[152] https://www.vupune.ac.in/

[153] https://www.vupune.ac.in/professional-courses/2-uncategorised/51-water-and-waste-water-treatment

[154] https://www.vupune.ac.in/water-quality-centre-of-excellence

[155] Karne, H. U.; Bhatkhande, D.; Jabade, S. Mesophilic and thermophilic anaerobic digestion of faecal sludge in a pilot plant digester. *Int. J. Environ. Studies* **2018**, *75*, 484-495.

[156] Khamparia, S.; Jaspal, D. K. Xanthium strumarium L. seed hull as a zero cost alternative for Rhodamine B dye removal. *J. Environ. Management* **2017**, *197*, 498-506.

[157] http://www.wilo-foundation.de/en.html

[158] https://www.tudelft.nl/en/ceg/research/stories-of-science/water-treatment-in-india-a-matter-for-the-local-community/

[159] Ghasimi, D. S. M.; Tao, Y.; de Kreuk, M.; van Lier, J. B.; Abbas, B.; Zandvoort, M. H. Digester performance and microbial community changes in thermophilic and mesophilic sequencing batch reactors fed with the fine sieved fraction of municipal sewage. *Water Res.* **2015**, *87*, 483-493.

[160] Ghasimi, D. S. M.; Zandvoort, M. H.; Adriaanse, M.; van Lier, J. B.; de Kreuk, M. Comparative analysis of the digestibility of sewage fine sieved fraction and hygiene paper produced from virgin fibers and recycled fibers. *Waste Management* **2016**, *53*, 156-164.

[161] Lindeboom, R. E. F.; Ilgrande, C.; Carvajal-Arroyo, J. M.; Coninx, I.; Van Hoey, O.; Roume, H.; Morozova, J.; Udert, K. M.; Sas, B.; Paille, C.; Lasseur, C.; Ilyin, V.; Clauwaert, P.; Leys, N.; Vlaeminck, S. E. Nitrogen cycle microorganisms can be reactivated after Space exposure. *Sci. Reports* **2018,** *8*, 1-7.

[162] https://www.waterworld.com/international/article/16208011/the-top-global-water-research-institutes

[163] https://www.susana.org/en/community/partners/list/details/65

[164] https://www.tuhh.de/alt/aww/research.html

[165] Meyer, W.; Reich, M.; Beier, S.; Behrendt, J.; Gulyas, H.; Otterpohl, R. Measured and predicted environmental concentrations of carbamazepine, diclofenac, and metoprolol in small and medium rivers in northern Germany. *Environ. Monitoring Assessment* **2016**, *188*, 1-16.

[166] Ramaswami, S.; Behrendt, J.; Otterpohl, R. Comparison of NF-RO and RO-NF for the treatment of mature landfill leachates: a guide for landfill operators. *Membranes* **2018**, *8*, 17/1-17/13.

[167] https://mof.gov.il/en/InternationalAffairs/InfrastructuresAndProjects/Projects/Pages/Background_DesalinationInIsrael.aspx

[168] https://www.ide-tech.com/en/

[169] https://www.water-technology.net/projects/sorek-desalination-plant/

[170] https://innovationisrael.org.il/en/page/calls-proposals

[171] https://www.isf.org.il/

[172] https://www.gov.il/en/departments/ministry_of_science_and_technology

[173] https://in.bgu.ac.il/en/bidr/ziwr/Pages/default.aspx

[174] https://in.bgu.ac.il/en/bidr/ziwr/ehm/Pages/staff/gross.aspx

[175] http://departments.agri.huji.ac.il/soils/index.shtml

[176] https://soilandwater.agri.huji.ac.il/yaelmishael

[177] (a) Gardi, I.; Mishael, Y. G. Designing a regenerable stimuli-responsive grafted polymer-clay sorbent for filtration of water pollutants. *Sci. Technol. Adv. Mater.* **2018**, *19*, 588-598; (b) Levy, L.; Izbitski, A.; Mishael, Y. G. Enhanced gemfibrozil removal from treated wastewater by designed "loopy" clay-polycation sorbents: Effect of diclofenac and effluent organic matter. *Appl. Clay Sci.* **2019**, *182*, 105278.

[178] https://chemistry.technion.ac.il/grand-water-research-institute-gwri/

[179] https://cee.technion.ac.il/members/ori-lahav/

[180] Barba, D.; Caputi, P.; Cifoni, D. Drinking water supply of Italy. *Desalination* **1997**, *113*, 111-117.

[181] https://www.power-technology.com/projects/sarlux/

[182] https://www.ide-tech.com/wp-content/uploads/2018/06/Sarlux-Italy-Project.pdf

[183] https://www.ide-tech.com/en/

[184] https://www.acciona-agua.com/pressroom/news/2018/january/acciona-builds-largest-desalination-plant-industrial-use-mediterranean-sardinia-italy/

[185] https://www.environmental-expert.com/companies/european-desalination-society-eds-24050

[186] Casimiro, S.; Ahmed, M. K. A.; Cardoso, J. P.; Mendes, J. F. Reverse osmosis powered by concentrating solar power (CSP): a case study for Trapani, Sicily. *Desalination Water Treatment* **2017**, *61*, 183-195.

[187] Lo Mastro, F.; Mistretta, M. Thermoeconomic analysis of a coupled municipal solid waste thermovalorization-MSF desalination plant: an Italian case study. *Desalination* **2006**, *196*, 293-305.

[188] Cardona, E.; Culotta, S.; Mangiaracina, L.; Terranova, R. Multi-flash desalination and cogeneration to solve the water supply problems in southeastern Sicily. *Desalination* **1996**, *106*, 121-130.

[189] https://www.membrane.it/en/article/index/about-us-company

[190] https://www.unibo.it/sitoweb/v.ciriello/cv-en

[191] Felisa, G.; Ciriello, V.; Antonellini, M.; Di Federico, V.; Tartakovsky, D. M. Data-driven models of groundwater salinization in coastal plains. *J. Hydrology* **2015**, *531*, 187-197.

[192] Felisa, G.; Lauriola, I.; Pedrazzoli, P.; Di Federico, V.; Ciriello, V. Sustainability analysis of alternative long-term management strategies for water supply systems: a case study in Reggio Emilia (Italy). *Water* **2019**, *11*, 450.

[193] https://www.polimi.it/index.php?id=6098&L=1&sel_dipartimento=267165713

[194] Colombo, N.; Bocchiola, D.; Martin, M.; Confortola, G.; Salerno, F.; Godone, D.; D'Amico, M. E.; Freppaz, M. High export of nitrogen and dissolved organic carbon from an Alpine glacier (Indren Glacier, NW Italian Alps). *Aquatic Sci.* **2019**, *81*, 1-13.

[195] Becciu, G.; Paoletti, A. Random characteristics of runoff coefficient in urban catchments. *Water Sci. Technol.* **1997**, *36*, 39-44.

[196] https://www.jetro.go.jp/en/mjcompany/kyowa.html

[197] https://www.sbmc.or.jp/english/060727/Sea_Water_Desalination_Plant.html

[198] https://www.toyobo-global.com/seihin/hq/

[199] https://www.nitto.com/jp/en/

[200] http://www.jwrc-net.or.jp/english/

[201] https://www.ce.t.kyoto-u.ac.jp/en/information/laboratory/water_resources

[202] http://www2.hyd.ncku.edu.tw/iche2016/PDF/0.%20Keynote%20Lectures/Keynote%20Lectures-03.pdf

[203] https://www.researchgate.net/profile/Yutaka_Ichikawa

[204] https://www.researchgate.net/scientific-contributions/2008467877_Kazuaki_Yorozu

[205] http://www.ues.tmu.ac.jp/civil/english/contents/HydrologyandWaterResources.html

[206] Nguyen, T. T.; Kawamura, A.; Tong, T. N.; Nakagawa, N.; Amaguchi, H.; Gilbuena, R. Clustering spatio-seasonal hydrogeochemical data using self-organizing maps for groundwater quality assessment in the Red River Delta, Vietnam. *J. Hydrology* **2015**, *522*, 661-673.

[207] Jin, Y.-H.; Kawamura, A.; Park, S.-C.; Nakagawa, N.; Amaguchi, H.; Olsson, J. Spatiotemporal classification of environmental monitoring data in the Yeongsan River basin, Korea, using self-organizing maps. *J. Environ. Monitoring* **2011**, *13*, 2886-2894.

[208] https://www.kobe-u.ac.jp/en/index.html

[209] http://www.research.kobe-u.ac.jp/ans-water/index_e.html

[210] https://www.researchgate.net/profile/Tanakamaru_Haruya

[211] https://www.researchgate.net/profile/Akio_Tada

[212] https://smartwatermagazine.com/specials/desalination-saudi-arabia

[213] http://www.kfupm.edu.sa/default.aspx

[214] http://www.kfupm.edu.sa/departments/ce/Pages/en/Water-Resources-and-Environmental-Engineering-Lab.aspx

[215] http://www.kfupm.edu.sa/departments/chem/Pages/en/Polymer,-Materials-And-Nanoscience.aspx

[216] https://faculty.kfupm.edu.sa/CHEM/shaikh/

[217] https://faculty.kfupm.edu.sa/CHEM/othmanc/

[218] https://kaust.edu.sa/en

[219] https://wdrc.kaust.edu.sa/Pages/Home.aspx

[220] https://wdrc.kaust.edu.sa/Pages/Vrouwenvelder.aspx

[221] https://wdrc.kaust.edu.sa/Pages/McCabe.aspx

[222] https://www.acciona.com/

[223] https://www.acciona.com/pressroom/news/2020/april/acciona-will-build-its-fourth-desalination-plant-in-saudi-arabia/

[224] http://www.abengoa.com/web/en/

[225] https://www.ediweekly.com/saudi-solar-powered-desalination-plant-will-worlds-largest/

[226] https://www.swcc.gov.sa/english/Pages/Home.aspx

[227] https://www.worldometers.info/world-population/south-korea-population/

[228] http://www.wepa-db.net/policies/state/southkorea/southkorea.htm

[229] https://www.waterworld.com/international/desalination/article/16201474/south-korea

[230] https://www.sciencedaily.com/releases/2018/07/180712100519.htm

[231] https://en.snu.ac.kr/

[232] https://www.researchgate.net/profile/Mooyoung_Han

[233] https://ace.korea.edu/ace_en/about/overview.do

[234] http://korea.edu/mbshome/mbs/en/index.do

[235] https://scholar.google.com/citations?user=c_iisoEAAAAJ&hl=en

[236] https://www.researchgate.net/profile/Chulsang_Yoo

[237] https://www.poscoenc.com:446/eng/about/introduction.asp

[238] https://www.poscoenc.com:446/eng/business/environment_intro.asp

[239] https://www.water-technology.net/news/newsdecentralised-water-supply-4431815/

[240] Woo, S. W.; Park, B. S.; Lee, W. N.; Park, Y. H.; Min, J. H.; Park, S. W.; You, S. N.; Jun, G. J.; Baek, Y. J. Seawater Intake System in Test Bed Seawater Reverse Osmosis (SWRO) Project. *Desalination Water Treatment* **2013**, *51*, 6238-6247.

[241] http://www.roplant.org/contents.asp?Depth1=7&Depth2=3&BoardAction=View&b_Seq=8704

[242] https://www.ajudaily.com/view/20190218163103055

[243] https://www.desaldata.com/blog/desaldata-weekly-february-22-2019

[244] http://english.kwater.or.kr/eng/about/sub02/kwaterPage.do?s_mid=1099

[245] Choi, J.-S.; Lee, S.; Kim, J.-M.; Choi, S. Small-Scale Desalination Plants in Korea: Technical Challenges. *Desalination* **2009**, *247*, 222-232.

[246] https://www.oecd.org/switzerland/Water-Resources-Allocation-Switzerland.pdf

[247] https://www.unine.ch/

[248] https://www.unine.ch/chyn/en/home/laboratoires.html

[249] https://www.unine.ch/chyn/home/collaborateurstrices/professeurs/hunkeler-daniel.html

[250] https://www.eawag.ch/en/

[251] https://ethz.ch/en/the-eth-zurich/organisation/eth-domain.html

[252] https://www.universitypositions.eu/university/eawag-swiss-federal-institute-of-aquatic-science-and-technology

[253] https://www.eawag.ch/en/research/humanwelfare/drinkingwater/

[254] https://www.eawag.ch/en/research/humanwelfare/wastewater/

[255] https://www.eawag.ch/en/research/water-for-ecosystem/pollutants/

[256] https://www.pureaqua.com/reverse-osmosis-water-treatment-in-switzerland/

[257] https://www.trunzwatersystems.com/

[258] https://filopur.com/

[259] https://www.ircwash.org/sites/default/files/community_water_supply_in_switzerland_0.pdf

[260] https://www.unece.org/fileadmin/DAM/env/water/Protocol_reports/reports_pdf_web/Switzerland_summary_report_en.pdf

[261] https://www.cardiff.ac.uk/water-research-institute/research/research-themes

[262] https://www.cardiff.ac.uk/people/view/541992-hubbard-christopher

[263] https://www.cranfield.ac.uk/

[264] https://www.cranfield.ac.uk/centres/water-sciences-institute

[265] https://www.cranfield.ac.uk/case-studies/research-case-studies/nano-membrane-toilet

[266] https://www.cranfield.ac.uk/research-projects/towards-more-effective-river-restoration-design

[267] https://www.cranfield.ac.uk/research-projects/sustaining-himalayan-water-resources-in-a-changing-climate

[268] https://www.cranfield.ac.uk/people/professor-paul-jeffrey-318515

[269] https://www.bath.ac.uk/

[270] https://www.bath.ac.uk/research-centres/water-innovation-and-research-centre-wirc-bath/

[271] https://www.theguardian.com/environment/2010/jun/02/thames-water-desalination-plant

[272] https://www.interserve.com/

[273] https://www.atkinsglobal.com/en-GB

[274] https://www.acciona.com/

[275] https://www.thegreenage.co.uk/cos/beckton-desalination-plant/

[276] https://www.pureaqua.com/reverse-osmosis-water-treatment-in-united-kingdom/

[277] https://www.lubron.co.uk/

[278] https://www.osmiowater.co.uk/contacts/

Made in United States
North Haven, CT
04 November 2026

21743225R00057